Australasian Marsupials and Monotremes

An Action Plan for their Conservation

Compiled by Michael Kennedy
IUCN/SSC Australasian Marsupial and Monotreme
Specialist Group

SPECIES SURVIVAL COMMISSION

IUCN

WWF

Australian National Parks and Wildlife Service

Sultanate of Oman

Contents

Foreword

The Australasian region is probably best known to people from other countries for its marsupials, especially kangaroos and the koala, and for the world's only egg-laying mammals, the platypus and the echidnas.

In recent decades, Australia has been criticised by people from other countries for allowing the harvesting of some large species of kangaroo, and there have been fact-finding visits by politicians, officials, and conservationists from Europe and north America.

Paradoxically, as this Action Plan shows, many of the large kangaroos have increased in abundance over the past 200 years, while, during the same period, many other marsupial species have become extinct, endangered, or vulnerable without similar interest being generated from other countries. Sadly, Australia has the worst record of any nation for mammal extinctions, even though much of the country appears little-altered by the impacts of humans and their technology.

Many of the marsupials and monotremes covered by this Action Plan are found beyond the political borders of Australia, occurring in Papua New Guinea and Indonesia, and in the case of one species, in the Solomon Islands. Outside Australia there has not been the same rate of marsupial extinction or decline, but as pressures for development are rapidly increasing in these countries, leading in some places to extensive habitat destruction, habitat protection and other conservation measures are now urgently needed to ensure that species do not become extinct.

Action is now needed urgently to ensure the conservation of Australasia's marsupials and monotremes. This Action Plan, developed over almost three years by members of the Australasian Marsupial and Monotreme Specialist Group of IUCN's Species Survival Commission, provides a blueprint for the conservation of two unique groups of mammals. It is the Specialist Group's hope that it will be used by governments, non-government organisations and individuals.

Dr. Andrew Burbidge
Chairman, 1991-1993
Australasian Marsupials and Monotremes Specialist Group

Acknowledgements

The preparation of this Action Plan simply would not have been possible without the generous contributions of time and knowledge given by countless individuals in Indonesia, Papua New Guinea, and Australia. The compiler would particularly like to thank the following colleagues who responded so regularly to requests for help, often providing minute detail for the species recovery outlines and distribution maps contained in this document. Many others provided valuable technical and strategic advice, including a long list of individuals who spared time to talk to him during his Australian and Papua New Guinea travels:

Frank Antram, Robert Baker, Robert Begg, Adrian Bradley, Richard Braithwaite, Mike Braysher, Linda Broome, Peter Brown, Debbie Callister, Robert Carr, Frank Carrick, Graeme Caughley, Mike Chilcott, Per Christensen, Peter Christie, Rob Close, Peter Copley, Doug Crossman, Stuart Davey, D. Davies, John DeJose, Martin Denny, Chris Dickman, Joan Dixon, Liz Dovey, Murray Evans, Tim Flannery, Greg Fraser, Tony Friend, Gordon Friend, Jack Giles, Greg Gordon, Tom Grant, Lance Hill, Greg Hocking, Ric How, Linette Humphrey, Peter Hunnam, Iamo Ila, Henry Isa, Richard Jakob-Hoff, Chris Johnson, Ken Johnson, Paul Kelly, Peter Kendrick, Max King, Jack Kinnear, Tom Kirkpatrick, Karol Kisokau, Darrell Kitchener, Guy Kula, Peter Latz, Bill Laurance, Tanya Leary, Tony Lee, Leong Lim, David Lindenmeyer, Leyton Llewellyn, Bert Lobert, Geoff Lundie-Jenkins, Dan Lunney, Ian Mansergh, Roger Martin, Pip Masters, Gerry Maynes, Malcolm McFarlane, Norm McKenzie, Keiran McNamara, Jeff McNeely, Peter Menkhorst, Jim Menzies, Darryl Miller, Keith Morris, Steve Morton, Alan Newsome, Ray Nias, Tony Norton, David Papps, Mike Patchett, Rod Pearce, David Pearson, Barry Pittock, Bill Poole, Barbara Porter, Tony Press, Bob Prince, George Rabb, Dave Read, Barry Richardson, Tony Robinson, R. Ross, David Rounsevell, Mike Saxon, A. Scotts, John Seebeck, Geoff Sharman, Jeff Short, Andrew Smith, Ric Southgate, Mark Stafford-Smith, Gary Steer, Ron Strahan, Simon Stuart, Graham Suckling, Effendy Sumardja, Peter Tierney, Steve Van Dyck, Bob Warneke, Chris Watts, Rod Wells, Keith Williams, George Wilson, John Winter, Dedee Woodside, Pat Woolley.

Thanks particularly to all the photographers for providing the transparencies for the Action Plan; to Tim Flannery for permission to reproduce his distribution maps for the New Guinea species; and to Lee Roberts and Christine Lawrence for their efforts in preparing the artwork for species distribution maps. Thanks also to Jared Diamond and Dick Schodde for permission to reproduce their maps and tables on proposed reserves in New Guinea. The above list is far from exhaustive, and the compiler gives his apologies to any person who really should have been thanked.

Special mention and thanks must go to Andrew Burbidge, Graeme George, Lester Seri, Widodo Ramono, Stephen Nash, and Tracey Smith for all their original and hard work during the compilation of this Action Plan—without their efforts this Action Plan might never have been completed!

Thanks must also go to WWF Australia for financially supporting the first Action Plan developed in Australia, and to the Commonwealth Government's Endangered Species Program for providing further resources to permit the extension of the project to Papua New Guinea and Indonesia. Their support has been invaluable.

The compiler takes full responsibility for any errors, inaccuracies or omissions that may be contained in this Action Plan, particularly where it has been necessary to accommodate any conflicting points of view.

Introduction

Australia, Papua New Guinea, and eastern Indonesia—referred to as Australasia in this Action Plan—together share the all the monotremes (egg-laying mammals) of the world and all the marsupials (pouched mammals) except those of the Americas.

Monotremes are a distinctive group of mammals that are distinguished by reproducing via eggs which are laid and incubated, even though they suckle their young. Only three species of monotremes exist—the platypus of Australia and two species of echidna, one in Australia and two on the island of New Guinea.

Marsupials also differ from the widespread placental mammals in their mode of reproduction. The young are born after a very short period of gestation and, after making an amazing journey to the pouch, continue their development attached to a teat. Over the many millions of years since separation of the Australasian region from Gondwana, the marsupials have radiated into many distinct groups, including carnivores such as the thylacine, quolls, and dunnarts, omnivores such as bandicoots, and herbivores such as possums, gliders, the koala, and kangaroos.

Although three groups of terrestrial placental mammals are native to Australia and the island of New Guinea (native rodents, flying foxes, and insectivorous bats), all the larger species are marsupials and for most people they typify the mammal fauna. In the other islands of eastern Indonesia several other placental groups occur with the marsupials, rodents, and bats.

The Australasian Marsupial and Monotreme Specialist Group (AMMSG) is one of many different Specialist Groups that operate throughout the world under the auspices of the IUCN—The World Conservation Union's Species Survival Commission. The goals of the Specialist Groups centre on identifying threatened species and threatening processes, and devising, in cooperation with governments and non-government conservation organisations (NGOs), management strategies and priorities for species conservation. Specialist Groups have broad memberships, and usually include independent wildlife scientists as well as wildlife researchers and managers employed by government agencies charged with the responsibility of conserving wildlife. The principal tool of Specialist Groups is the preparation of Action Plans—documents that clearly indicate which species are threatened with extinction and that identify appropriate actions for their conservation.

There is an urgent need for concerted action to conserve the marsupials and monotremes of the Australasian region. Australia has the worst extinction rate for mammals of any continent or country and Papua New Guinea and eastern Indonesia are undergoing rapid development which, if not properly planned and controlled, could threaten the habitat of many marsupials, as well as other species. Realizing this, the AMMSG decided to prepare an Action Plan that could provide an overall perspective of the problems that confront conservation agencies and NGOs in the region and recommend actions required before the year 2000.

In 1987, the then AMMSG chairperson, Dr. George Wilson, in conjunction with the compiler, approached the World Wide Fund for Nature (WWF) Australia for the resources to begin the action planning process in Australia. The application was favourably received by WWF and by early 1989 the project began with the compiler working on a part-time basis. Later in 1989, WWF and the AMMSG made an application to the Australian federal government's Endangered Species Advisory Committee for funds to permit the coverage of all Australasian marsupials and monotremes. This was also successful and by the end of 1989 the full project was underway.

The Action Plan has been drawn up to provide actions for the period 1991 to 2000, with a review to be undertaken in 1996. The Plan is primarily aimed at governments and their wildlife conservation and funding agencies, but is also targeted at non-government scientific, environmental, educational, and funding bodies and foundations. The AMMSG believes, however, that the Action Plan can be particularly useful to any individual or organization that is concerned with the conservation of the animals of the Australasian region.

The Action Plan firstly describes the zoogeography of Australasian marsupials and monotremes and the taxonomy used. It then provides a conservation overview on a country by country basis, dealing first with eastern Indonesia, then Papua New Guinea, and finally Australia. It then briefly discusses the management for conservation of small populations. The "Action" section of the Plan lists all species known to occur in the region, including information on their conservation status, and then provides information on a "Reserve List" of species that require monitoring to ensure that they do not enter one of the more critical threatened categories. Finally, detailed recommendations for each country are given. For each Australian species listed as Endangered or Vulnerable, a species recovery outline is provided in an Appendix. Importantly, the outline provides an estimate of the cost of each action.

This Action Plan lists 39 Endangered and Vulnerable species and 55 "Potentially Vulnerable" species (94 species of marsupials and two monotremes), that require some form of conservation action. Three Potentially Vulnerable species will remain listed as Rare in the IUCN Red List of Threatened Animals and the AMMSG will await the determination of the new IUCN conservation categories before deciding whether any other of the Potentially Vulnerable species require an entry in the Red List.

It should be noted that the authors of the sections on eastern Indonesia and Papua New Guinea report that there are many species that live outside protected areas and which have restricted geographic ranges or exist mainly in areas of high human population and that these should be considered to be threatened to some degree.

Priorities for Action

The AMMSG members have not assigned priorities within the Endangered and Vulnerable Species categories but clearly the Endangered category must receive attention before the Vulnerable species. Potentially Vulnerable species require careful monitoring to detect any further deterioration in their conservation status.

In Indonesia and Papua New Guinea the 12 species of Endangered and Vulnerable marsupials and monotremes must receive priority attention and the general lack of knowledge about many other species in these countries must be rectified by systematic biological survey.

In Australia there are many species that are already the subject of development or implementation of recovery plans or some other form of management or research. This Action Plan will provide, therefore, a useful review of activities from which gaps will be identified and priorities can be properly oriented. Indeed even before its completion the findings of this Plan were used by the Federal Government's Endangered Species Program in determining, with State and Territory Governments, priorities at a national level.

Implementation of recommendations over the 10 year life of this Plan will cost an estimated A$24 million in Australia alone. This figure excludes the rodent, bat, bird, reptile, frog, invertebrate, and plant groups in which there are also many threatened species that will add further cost.

Evaluation of Programs

No conservation plan should be implemented without attendant programs for performance monitoring and evaluation of management outcomes. There are considerable problems facing Indonesia (especially Irian Jaya) and Papua New Guinea but the long-term process of establishing a satisfactory system of protected areas will provide important markers for future evaluation. Papua New Guinea has initiated a structured review of conservation policy and institutional arrangements that will facilitate the evaluation of conservation programs. The World Bank and other government and non-government funding agencies are providing substantial resources for conservation in this country and IUCN, WWF and others will encourage the implementation of this Action Plan within the limits of these resources.

The AMMSG will continue to monitor and assess the implementation of the recommendations of this Action Plan throughout Australasia and will conduct a major review of achievements and directions in 1996.

The Action Plan in Action

It was especially encouraging to the AMMSG that Federal, State and Territory governments in Australia have already utilised recommendations within a draft of this Action Plan for developing joint recovery plans. This provides direct evidence of the beneficial role of Action Plans and Specialist Groups. Table 20, Chapter 4 shows the marsupials currently receiving support under this joint program.

WWF Australia is supporting many recovery programs and it was a major sponsor of this Action Plan. A further conservation program with emphasis on species conservation is being developed by WWF to assist the Papua New Guinea government. It also intends to produce an executive summary of this Action Plan for wide and regional distribution.

Threatened species legislation, a further recommendation of this Action Plan, has been passed or is in preparation by the Federal and several State governments. Action Plans for rodents and bats (recommended by this Plan), birds, reptiles, frogs, freshwater fish, non-marine invertebrates, and non-vascular plants have been commissioned by the federal government and will be published in 1992. Drafts of this document have been used by individuals and organisations working on species management and as evidence in several inquiries into conservation issues in Australia. Finally, the SSC is arranging for the preparation of an Action Plan for the marsupials of the Americas.

Altogether there has been a very pleasing start to the implementation of this Action Plan for Australasian marsupials and monotremes.

The Action Planning Process

This Action Plan was developed over a two-and-a-half year period. The initial stages were guided by the Action Plan Steering Committee which organised a regional workshop of the AMMSG. The workshop was attended by representatives from Papua New Guinea, non-government organisations, and the federal, state, and territory wildlife agencies. Indonesia was unable to provide input at this time. The meeting reviewed and developed the Action Plan recommendations and established the conservation status of most marsupials and monotremes within the region.

Data were gathered primarily through the wide circulation of status reviews, species recovery summaries, and draft texts to Specialist Group members and other relevant organisations and individuals. An editing team produced the core of the Action Plan text for Australia and country reports were received from Indonesia and Papua New Guinea. Lack of information on threatened New Guinea species prevented compilation of "species recovery outline" sheets for this area, although brief details on current threats and proposed conservation actions have been included.

Wide circulation of draft texts and species recovery outlines continued during 1990 and the difficult task of determining the conservation status of New Guinea species was completed. Additional text was received from AMMSG members following the IUCN General Assembly at Perth in late 1990. A final editing process was completed by a small number of expert volunteers in collaboration with the SSC at the IUCN headquarters in Switzerland. Information in this Action Plan was accurate at July, 1991.

Michael Kennedy
Action Plan Compiler and
AMMSG Secretary

1. Zoogeography and Taxonomy

Zoogeography of Australasian Marsupials and Monotremes

The marsupials of the Australasian Region are distributed across four national jurisdictions—those of Indonesia, Papua New Guinea, Australia, and the Solomon Islands. The monotremes are more restricted, occurring in Australia (the Australian mainland and Tasmania), Papua New Guinea (eastern New Guinea), and the Indonesian Province of Irian Jaya (western New Guinea).

The marsupials are numerically most abundant and diverse in Australia, with 142 species distributed among 5 orders, 16 families, and 49 genera. New Guinea also has a diverse marsupial fauna with 68 species in 3 orders, 9 families, and 26 genera spread over the main island and associated islands of the Sahul Shelf. In contrast, the eastern islands of the Indonesian archipelago west of the Sahul Shelf (Sulawesi, Maluku, Timor, and Sundas), have only 8 species in 2 orders, 3 families, and 6 genera. These islands form a transition zone, known as Wallacea, between the Oriental and Australasian Regions. The islands of Papua New Guinea (Admiralties, Bismarck Archipelago, etc.) have only four species in 2 orders, 3 families, and 4 genera. The Solomon Islands have only a single species of marsupial (*Phalanger orientalis*). As there is no evidence that the Solomon Island populations of this widespread species are threatened (T. Leary pers. comm.), there are no specific recommendations for the Solomon Islands in this Action Plan.

Marsupials do not naturally occur in New Zealand, although several species have been introduced there and many of these have established feral populations.

There are only three monotypic species of monotreme—the short-beaked echidna (*Tachyglossus aculeatus*) of Australia and New Guinea and the long-beaked echidna (*Zaglossus bruijni*) of New Guinea, both in the family Tachyglossidae; and the platypus (*Ornithorhynchus anatinus*) of Australia in the family Ornithorhynchidae.

The distinctiveness of the Australian mammal fauna is due to its long isolation from the faunas of other land masses since separation from the supercontinent of Gondwana early in the Tertiary, some 45 million years ago. The three present-day land masses of Tasmania, the Australian mainland and New Guinea have subsequently shared a long, common geological history and their respective mammal faunas reflect this. Their current separation by the marine barriers of Bass and Torres Straits is a very recent event, brought about by the rise in sea levels after the last Pleistocene ice age ended about 10,000 years ago. The dispersal of the marsupials beyond New Guinea towards the Pacific in the northeast and Asia to the northwest during the orogenic events of the late Tertiary has been of a fortuitous nature because of the more persistent marine barriers present off the Australian continental plate during this period.

Northern bettong, *Bettongia tropica* (photo by G. Gordon, Queensland National Parks and Wildlife Service/WWF Australia).

The rise in sea level that formed the Arafura Sea over the Sahul Shelf, culminating in the inundation of Torres Strait about 8,000 years ago, divided populations of both rain-forest and tropical savannah species. This left outliers of "Australian" species in southern New Guinea and of "New Guinea" species in the rain-forests of Cape York and the Aru Islands. Genetic divergence of these populations, which may have commenced long before inundation, has been sufficient for the recognition of subspecies in several cases. Where the conservation status of these outlier populations differs from the main population they are treated as separate entities for the purposes of this Action Plan, to emphasise national responsibilities.

The post-glacial rises in sea level also isolated many marsupials on continental Australian islands. In many cases, mainland populations have drastically declined or become extinct since European settlement of the continent, and these islands have become fortuitous refuges.

Of the mountainous island groups off the northwest and southeast coasts of New Guinea, endemic species of marsupials occur on Waigeo, Goodenough, Fergusson, Kiriwina, and Woodlark, reflecting longer periods of isolation from their congeners on the mainland. All of these must be considered vulnerable to overexploitation and loss of habitat.

Families of Marsupials and Monotremes in Australasia

The following is a brief description of the characteristics of each of the families that make up modern day marsupials and monotremes:

Monotremes

Mammals which lay soft-shelled eggs, hatch their young in a temporary abdominal pouch, and suckle their young from pores in the skin over the mammary gland.

Tachyglossidae—Echidnas
Monotremes in which the snout has been modified to form an elongated beak-like structure. Teeth are absent and the tongue is long and protrusible, adapted for the ingestion of ants and termites (*Tachyglossus*) or earthworms (*Zaglossus*). Some hairs are modified to form sharp spines. Both monotypic genera are terrestrial and sub-fossorial. *Zaglossus* is confined to New Guinea and is Vulnerable.

Ornithorhynchidae—Platypus
A small amphibious monotreme with the snout modified to form a pliable duck-like bill used to sift organisms from stream sediments. Adults are toothless, using horny plates to grind food. Webs on the feet used in swimming are folded back for burrowing. The bill is highly innervated and animals respond to electrical fields in the water, possibly using this faculty to locate food and avoid objects underwater. Males have a venomous spur on the hind limb. One extant species, Potentially Vulnerable.

Marsupials

Therian mammals bearing live embryonic young that undertake their initial growth attached to abdominal teats, often, but not always, protected by a pouch. Greater diversity at the higher taxonomic levels is now recognised among the Australasian marsupials following cladistic analysis of new molecular and morphological data, including much new material from fossil deposits. The "possums" are no longer believed to be monophyletic.

Thylacinidae—Thylacine
The largest carnivorous marsupial to have survived into modern times. Convergently dog-like, but with relatively short legs and a stiff tail. The pouch opens to the rear. Common and regarded as a pest in Tasmania at the time of European settlement, the last known specimen died in captivity in 1936. Officially extinct, but there are frequent unconfirmed reports of its continued existence in northern Tasmania.

Dasyuridae—Dasyurids
A large family of carnivorous and insectivorous marsupials, primarily terrestrial and nocturnal, but with some species arboreal and active during the day. Taxonomy is unstable, but at least 16 genera are recognised in five sub-families, ranging in size from the shrew-like *Planigale* species (4 g) to the heavily-built scavenger *Sarcophilus* (8 kg). Lifespans of most species are relatively short, with high population turnovers. Eighteen species are considered to be at some degree of risk.

Myrmecobiidae—Numbat
A small terrestrial diurnal marsupial adapted to feed exclusively on termites. Utilises burrows and hollow logs for shelter. Once distributed over 25% of the continent, it is now confined to small remnant patches of undisturbed habitat in the southwest of Western Australia. Endangered.

Peramelidae—Australian Bandicoots and Bilbies
Bandicoots combine the polyprotodont dentition of the Dasyuroidea with the syndactylous hind foot of the diprotodont marsupials. The peramelids are small terrestrial omnivorous marsupials with elongated snouts, stocky bodies, and short limbs, ranging in size from 400 g to 4.7 kg. They obtain most of their food by scratching in the soil with their forefeet. The peramelids apparently represent an Australian radiation, with only one species extending into the savannahs of southern New Guinea. They have suffered extensively from European settlement—one genus and three species are presumed extinct and a further five species are threatened.

Peroryctidae—Rain-forest Bandicoots
A family of bandicoots that has apparently radiated in the rain forests of the northern Australian land mass, now largely confined

Northern hairy-nosed wombat, *Lasiorhinus krefftii* (photo by Jiri Lochman/WWF Australia).

to New Guinea, and distinguished from peramelids on skull characteristics. The size range is greater, from 100 g to over 5 kg. One genus and seven species are threatened.

Notoryctidae—Marsupial Mole
A small fossorial marsupial of sandy deserts, with reduced limbs, no ear pinnae or external evidence of eyes, convergently similar to placental moles, and feeding primarily on insect eggs, larvae, and pupae. Affinities with other marsupials are not clear. In the absence of good data it is regarded as Potentially Vulnerable.

Phascolarctidae—Koala
A medium-sized diprotodont marsupial (7-13 kg), adapted to feeding on foliage of a limited range of *Eucalyptus* species. The only extant species of its family, it is secondarily adapted for arboreal life, virtually tail-less, with a rear opening pouch, and a manus which is strongly clawed and forcipate—digits I and II oppose the other three. Local populations face extinction from habitat loss and disease, but the species is classified as Potentially Vulnerable.

Vombatidae—Wombats
Medium to large (up to 35 kg) heavily-built herbivorous terrestrial marsupials, with continuously growing open-rooted molars for grazing, building burrows for shelter. There are three extant species in two genera, the remnants of a more extensive Tertiary radiation (seven genera). One species is Endangered.

Wombats and koalas are the remnants of a more extensive marsupial radiation that included the rhinoceros-sized diprotodontids and the predatory thylacoleontids, that died out in the Pleistocene.

Phalangeridae—Cuscuses and Brushtail Possums
Medium sized (1-6.5 kg) omnivorous to herbivorous marsupials, predominantly arboreal, with prehensile tails and granulated foot pads. Taxonomy is unstable but there are at least 21 species in six genera. Most are tropical rain-forest species. Fourteen are in need of conservation action.

Potoroidae—Potoroos and Bettongs
Small (360 g-3.5 kg) macropodoid marsupials with short fore-limbs and well developed hindlimbs, these are terrestrial omnivores with the Potoroinae feeding extensively on hypogeal fungi and *Hypsiprymnodon* partly frugivorous. There are ten modern species in five genera, confined to Australia. One modern genus and two species are extinct, and all eight extant species are under some degree of threat. The family was much more diverse in the Tertiary.

Macropodidae—Kangaroos and Wallabies
A large family of bipedal herbivorous marsupials with 54 species in 11 genera, ranging in size from the hare-wallabies at 1 kg to the red and grey kangaroos that may weigh 80 kg and stand 2 m tall. The diverse family covers all habitats in Australia and New Guinea and includes rockpile and cliff-dwelling rock-wallabies (*Petrogale*) and secondarily-arboreal tree-kangaroos (*Dendrolagus*). Dental adaptations have evolved to support either grazing or browsing feeding modes. Four modern species are extinct, while one genus and 22 species are threatened to some degree.

Proserpine rock-wallaby, *Petrogale persephone* (photo by J.R.D. Bartlett).

Burramyidae—Pygmy-possums
Very small scansorial to arboreal marsupials, 10-60 g in weight with long prehensile tails. These omnivores feed on insects, seeds, soft fruit, nectar, and pollen. There are five species in two genera, most in Australia with one species shared with New Guinea. The monotypic *Burramys* is Endangered.

Pseudocheiridae—Ringtail Possums
Small (100 g-2.2 kg) arboreal folivorous marsupials with prehensile tails, selenodont molars, and a forcipate manus. One monotypic genus (*Petauroides*) has a gliding membrane between fore- and hind-limbs. It is distributed through forested regions in Australia and New Guinea. Two Australian species are threatened, two New Guinea species are threatened, and one is of unknown status.

Petauridae—Gliders and Striped Possums
Small (70-700 g) arboreal omnivorous marsupials feeding on invertebrates, nectar, and plant exudates. *Petaurus* has gliding membranes between fore- and hind-limbs. One genus and six species are threatened in Australia and New Guinea. One Australian species is of unknown status.

Tarsipedidae—Honey-possum
A very small specialised nectarivorous marsupial with a brush-tipped tongue, most closely related to the acrobatids. Confined to the southwestern corner of Western Australia, where it is locally common. Regarded as Potentially Vulnerable.

Acrobatidae—Feather-tailed Possums
Very small arboreal omnivorous marsupials distinguished from other families by the unique structure of their inner ear. Both monotypic genera have a tail laterally fringed with stiff hairs. *Acrobates* in Australia has a gliding membrane. *Distoechurus* is restricted to New Guinea. Both species are common.

Taxonomy of Marsupials and Monotremes reviewed in the Action Plan

The taxonomy used in this Action Plan follows Walton (1988) for Australian genera and species. Aplin and Archer (1987) are followed at higher taxonomic levels. A list compiled by George (1992-Appendix 1), updating the checklist of Laurie and Hill (1954), is followed for species from eastern Indonesia, Papua New Guinea, and the Solomon Islands. Several species of marsupial have been described recently and more await description.

Variations from the taxonomy of Walton are as follows:

1. Species of *Parantechinus* and *Dasycercus*, which have been allocated by some authors to genera other than those recognised by Walton (1988), are shown at sub-generic level to indicate their affinities.
2. Two Australian species described since the Zoological Catalogue went to press, *Pseudantechinus ningbing* Kitchener, 1988 and *Pseudantechinus woolleyae* Kitchener & Caputi, 1988, are placed in *Parantechinus* in accordance with Walton (1988). *Pseudantechinus mimulus* (Thomas 1906), recently re-described and re-instated as a species by Kitchener (1991) is also here included in *Parantechinus*.
3. Where Australian species are shared with the island of New Guinea, the taxonomic authorities listed in George (Appendix 1) are followed, e.g. the bandicoot family Peroryctidae is recognised following Groves and Flannery (1990).
4. *Pseudocheirus occidentalis* Thomas, 1888, the western ringtail, is here recognised as a species, separate from *P. peregrinus*. Whatever its taxonomic status, its conservation status should be separately assessed because of its geographical isolation.
5. *Petaurus gracilis* (de Vis, 1883) is tentatively separated from *P. norfolcensis* pending clarification of its taxonomic status. This taxon was rediscovered too late for a conservation status to be provided in this Plan.

No conservation status is assigned in this plan to three species recognised in Walton (1988), pending resolution of their taxonomic status. These are:

1. *Sminthopsis fuliginosus* (Gould, 1852); type locality: King George Sound (Albany area), Western Australia.
2. *Isoodon arnhemensis* Lyne & Mort, 1981; type locality: Melville Bay, Cape Arnhem area, Northern Territory.
3. *Notoryctes caurinus* Thomas, 1920; type locality: Wallal, northwestern Western Australia.

With these variations, currently described Australasian monotremes and marsupials are listed in Chapter 6, Table 21.

2. Conservation of Marsupials and Monotremes in Indonesia

Widodo S. Ramono[1] and Stephen V. Nash[2]
(Based on the work of R. G. Petocz 1984)

Much of Irian Jaya, Indonesia's eastern-most province and the western half of the island of New Guinea, is still largely untouched—for now. However, large-scale mining, forestry, and resettlement projects threaten increasingly larger areas of habitat—hundreds of thousands of hectares in individual projects—and so perhaps a significant number of species can be considered threatened, until the Irian Jaya Protected Area System is accepted, endorsed, and implemented by the Indonesian government. Any Irian Jaya marsupial or monotreme with a limited geographic or altitudinal range should be considered Potentially Vulnerable, as change arrives quickly in Irian Jaya.

Of particular concern are those species restricted to Irian Jaya. Four of these occur in mountain forests: the red-bellied dasyure/Tikus Berkantung Doria (*Phascolosore doriae*), regarded by Flannery (1990) as "Uncommon"; the mouse bandicoot/Bandikut Tikus (*Microperoryctes murina*) regarded by Flannery (1990) and George & Maynes (1990) as "Rare"; the Weyland ringtail/Kuskus Ekor Kait (*Pseudocheirus caroli*), and the Arfak ringtail/Kuskus Ekor Kait Arfak (*P. schlegeli*), both also listed as "Rare" by Flannery. The three "Rare" species are known only from very few museum specimens (only the type specimen in the case of the Arfak ringtail). The endemic Waigeo spotted cuscus (*Spilocuscus papuensis*) and the black or white-throated tree-kangaroo/Kangguru Pohon (*Dendrolagus ursinus*) of the Cendrawasih (Vogelkop) Peninsula, must be considered Potentially Vulnerable because of their restricted geographical range and their size which makes them attractive game species.

Three other species of conservation concern in Irian Jaya are primarily western in their distribution, but with minor range extensions into Papua New Guinea. These are the great-tailed triok (*Dactylopsila megalura*), regarded by Flannery (1990) as "Rare"; the woolly or D'Albertis ringtail/Kuskus Ekor Kait Albertis (*Pseudochirops albertisii*), and the grizzled tree-kangaroo/Kangguru Pohon Coklat (*D. ursinus*), both regarded by Flannery as "Uncommon".

Perhaps worse off still are the species found in neighboring Maluku province and the islands of Timor, Wetar, Sulawesi, and others.

There are several endemic genera and species in these islands. The large Celebes cuscus (*Ailurops ursinus*) is placed by Flannery et al. (1987) in its own sub-family. Also found on Sulawesi is the diminutive little Celebes cuscus (*Strigocuscus celebensis*) and on the nearby Peleng Island group the Peleng Island cuscus (*Phalanger pelengensis*) occurs. Within Moluku Province there are at least three endemics—the Moluccan cuscus (*P. ornatus*) of the Halmaheira group, the Obi Island cuscus (*P. rothschildi*) and the Seram bandicoot (*Rhynchomeles prattorum*). More widespread through Wallacea are the common grey cuscus (*P. orientalis*) and the common spotted cuscus (*Spilocuscus maculatus*), but there are endemic subspecies within Indonesian territory which also require monitoring. There is little information on the current state of the environment in most of these areas, but studies of Wallacean birds have shown that natural vegetation on many smaller islands has been lost through agricultural development, bringing with it the loss of endemic wildlife. Any endemic marsupial species in these areas should be tagged for status surveys, or at the very least a categorization of existing habitats should be made.

Long-beaked echidna, *Zaglossus bruijni* (photo by T. Flannery).

[1]Sub-Directorate of Species Monitoring, Directorate General of Forest Protection and Nature Conservation, Republic of Indonesia.
[2]Coordinator, WWF Irian Jaya Conservation Program.

Table 1. Threatened Indonesian marsupials and monotremes.

Common Name	Scientific Name	Status
Long-beaked echidna	*Zaglossus bruijni*	Vulnerable
Red-cheeked dunnart	*Sminthopsis virginiae*	Potentially Vulnerable
Clara or dimorphic bandicoot	*Echymipera clara*	Potentially Vulnerable
Mouse bandicoot	*Microperoryctes murina*	Potentially Vulnerable
Seram bandicoot	*Rhynchomeles prattorum*	Vulnerable
Large Celebes cuscus	*Ailurops ursinus*	Potentially Vulnerable
Aru Island ground cuscus	*Phalanger g. gymnotis*	Potentially Vulnerable
Moluccan cuscus	*Phalanger ornatus*	Potentially Vulnerable
Peleng Island cuscus	*Phalanger pelengensis*	Potentially Vulnerable
Obi Island cuscus	*Phalanger rothschildi*	Potentially Vulnerable
Stein's cuscus	*Phalanger vestitus*	Potentially Vulnerable
Waigeo spotted cuscus	*Spilocuscus papuensis*	Potentially Vulnerable
Black-spotted cuscus	*Spilocuscus rufoniger*	Vulnerable
Little Celebes cuscus	*Strigocuscus celebensis*	Potentially Vulnerable
Doria's tree-kangaroo	*Dendrolagus dorianus*	Vulnerable
Goodfellow's tree-kangaroo	*Dendrolagus goodfellowi*	Vulnerable
White-throated tree-kangaroo	*Dendrolagus ursinus*	Potentially Vulnerable
Grizzled tree-kangaroo	*Dendrolagus inustus*	Potentially Vulnerable
Great-tailed triok/Striped possum	*Dactylopsila megalura*	Vulnerable
Weyland ringtail	*Pseudocheirus caroli*	Potentially Vulnerable
Arfak ringtail	*Pseudocheirus schlegeli*	Potentially Vulnerable

There are few mammal distributional data outside existing or proposed protected areas, and all of these records and predicted occurrences are included in the following text and tables. The main point is that for the conservation of marsupials and monotremes to be possible in Irian Jaya, the reserve system must become a reality.

All Irian Jaya distribution records in this report were taken from Petocz and de Fretes (1983). Others are from van der Zon (1979) with additional information from G. George (pers. comm.). Table 1 lists the threatened Indonesian marsupials and monotremes identified by this Action Plan—six Vulnerable and 15 Potentially Vulnerable species.

The Irian Jaya Protected Area System (Adapted from Petocz 1984)

The Irian Jaya protected area system was designed between 1980 and 1983 by Dr. R. Petocz, and is based on the following considerations:

Habitat and Altitudinal Considerations

The very rich and diverse biota of Irian Jaya ranges over a vast continuum of habitats from the very lowest elevations at sea level upwards through the highest mountains in Southeast Asia (4,884 m). However, practically all of the thousands of wildlife species in the province are limited to a small part of this gradient. It is therefore of great importance that the entire altitudinal spectrum of the province be covered within the reserve design.

Of equal importance is the variation in habitat types that occurs within a narrow altitudinal span. The distribution of habitats in the horizontal sense is of equal importance to consider alongside those changes that occur with altitude. Likewise, habitats of a more limited distribution must be identified and covered within the reserve system design.

Endemism and Species Diversity

Another key element in the selection of reserves is an understanding of species distribution, in particular those species which are endemic to the province. For example, 50% of the endemic marsupials are montane species, so it is therefore essential to include all the major mountain ranges as an important part of the system design.

Biogeographical Considerations

The most prominent of the major biogeographic zones of Irian Jaya is the central cordillera, with no passes below 1500 m. It effectively prevents gene flow, isolates many hill and lowland species on either side of the range; and it defines two of the major lowland biogeographic units, the northern plains and foothills and the southern lowlands. The range itself is a formidable barrier to gene flow along its length. Less mobile montane elements have little gene flow along the backbone of the cordillera so that many populations develop new characteristics and visible morphological changes from west to east.

There are other distinctive biogeographical units. These include the lakes-plains depression of the upper Mamberamo and Rouffaer rivers; the dry southeast lowlands in the monsoonal plain; the isolated montane units of the Cyclops, Foja, and Van Rees mountains; the Tamrau and Arfak mountains; the Wandamen mountains; and the Fakfak and Kumawa mountains.

The off-shore islands fall into two biogeographical groups. The islands of the Sahul Shelf, which include Waigeo, Batanta, Salawati, and Misool to the west, and Yapen to the north, all had recent intermittent land connections with mainland New Guinea. Those which lie off the Sahul Shelf and had no connections with

New Guinea in the recent past are Kofiau in the far west, and Biak, Supiori, and Numfor to the north.

All of these biogeographical factors were considered in designing the protected area system.

Area Considerations and the Size of Reserves

The spatial requirements of the many different elements of Irian Jaya's fauna vary every bit as much as altitudinal range and habitat preference. To be meaningful, each reserve must be sufficiently large to protect self-sustaining viable populations of all elements of the biota within carefully selected biogeographical areas which cover the diversity and numbers in the province.

The protected area system as proposed by the World Wide Fund for Nature (WWF) now stands at over 50 areas (Tables 2 and 3 combined), covering some 82,076 square kilometers, or 19.8% of the province of Irian Jaya. The protected area system as it has been designed is expected to include every combination of habitats and altitudinal zones, and the full faunal and floral diversity of Irian Jaya.

Implementation of the Protected Area System

To date, implementation of the system has not occurred. Less than half the system's areas are formally recognised. Since the protected area system was presented in 1983, no further conservation areas have been gazetted (formally recognised). Table 2 lists those reserves that have been gazetted to date, and the year of the original decree.

Of the 51 areas recommended by WWF for the protected area system, 28 have not yet been recognised (Table 3). Of the 24 areas already gazetted, 23 are included in the protected area system design, but 18 of these require boundary and area changes to be effective. One major existing reserve, Enarotali Strict Nature Reserve, has been proposed for de-gazettal by WWF. Figure 1 shows the existing conservation areas in Irian Jaya.

Table 3. Proposed protected areas in Irian Jaya which require gazettal (excluding four turtle-nesting beaches and three proposed maritime reserves).

Conservation Area	Recommended Area (ha)
Mamberamo-Foja NP*	1,661,100
Weyland Mountains SNR	7,274,100
Pulau Numfor SNR	1,500
Arfak Mountains SNR	65,300
Teluk Bintuni SNR	200,000
Tamrau Utara SNR	331,300
Tamrau Selatan SNR	245,600
Pulau Kofiau SNR	2,000
Kumawa Mountains SNR	184,200
Fak fak Mountains SNR	191,600
Kumbe-Merauke SNR	100,600
Rouffaer River WS	531,100
Mingima WS	3,800
Wagura-Kote WS	18,200
Pulau Mapia WS	250
Pulau Sayang WS	2,468
Kepulauan Asia & Ayu WS	1,531
Seram River RP	1,000
Parieri RP	2,000
Klamono RP	10,000
Beriat Mountains RP	6,300

*Recommended as a World Natural Heritage Site

Table 2. Irian Jaya conservation areas already gazetted.

Area	Year	Gazetted Area	WWF Proposed Area
Teluk Yautefa RP	1978	1,650 ha	1,650 ha
Lorentz SNR	1978	2,150,000 ha	1,483,200 ha
Cyclops SNR	1978	22,520 ha	36,800 ha
Jayawijaya WS	1981	800,000 ha	942,000 ha
Nabire RP	1980	100 ha	100 ha
Enarotali SNR*	1980	300,000 ha	
Anggremeos WS	1981	2,500 ha	2,100 ha
Yapen Tengah SNR	1982	59,000 ha	79,700 ha
Inggresau WS	1982	300 ha	280 ha
Supiori SNR	1982	42,000 ha	40,800 ha
Biak Utara SNR	1982	11,000 ha	9,600 ha
Gunung Meja RP	1980	500 ha	500 ha
Wandamen SNR	1981	48,000 ha	90,600 ha
Sorong RP	1981	945 ha	945 ha
Batanta Barat SNR	1981	10,000 ha	10,000 ha
Waigeo Barat SNR	1981	153,000 ha	113,700 ha
Salawati Utara SNR	1982	57,000 ha	67,800 ha
Misool Selatan SNR	1982	84,000 ha	111,900 ha
Sabuda dan Tataruga	1980	450 ha	450 ha
Wasur WS	1978	600,000 ha	720,900 ha
Wasur Extension WS	1982	98,000 ha	
Rawa Biru WS	1978	4,000 ha	10,400 ha
Pulau Pombo SNR	1982	100 ha	100 ha
Danau Bian WS	1982	60,000 ha	106,700 ha

RP=Recreation Park; SNR=Strict Nature Reserve; WS= Wildlife Sanctuary; *= recommended by WWF for de-gazettal

The Protected Area System and Marsupials and Monotremes

Table 4 provides a list of threatened monotremes and marsupials found, expected, or which may occur within the protected area system of Irian Jaya. Asterisked species are confirmed records (from Petocz and de Frates 1983—additional information from G. George, pers. comm.).

Problems with the Protected Area System

The protected area system of Irian Jaya is a theoretical plan, one which has yet to be accepted by the Indonesian government. The reality is that there are serious land-use conflicts over areas proposed for conservation in Irian Jaya.

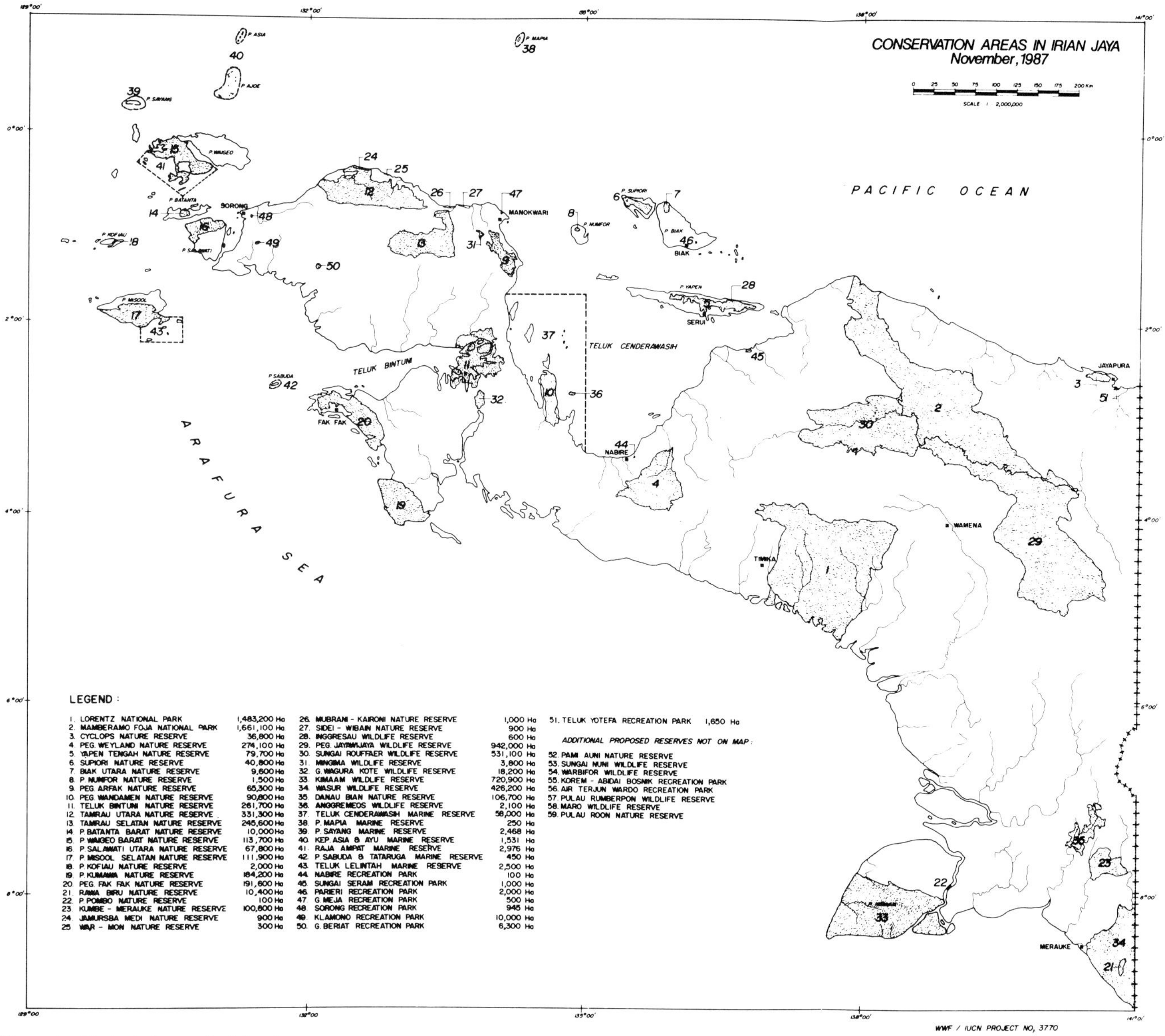

Figure 1. Protected areas in Irian Jaya.

Boundary conflicts

The boundaries suggested by WWF have been chosen primarily with biological considerations in mind. These serve as general guidelines in choosing and making boundaries on the ground. The future of conservation areas in Irian Jaya is dependent on the support and cooperation of local people living in and around the conservation areas. Boundaries must be chosen during the site-specific management planning process, with full involvement of local people. This community-participatory approach will ensure that boundaries do not conflict with present or predictable land use needs, and that boundaries and the boundary choosing process are compatible with the local traditional land management structures. This integration of local communities with the conservation authority in the management of boundary areas is the key to long-term conservation management in Irian Jaya. Unfortunately, the Department of Forestry's boundary marking section places boundaries with little regard to biological or human needs, resulting in inadequate conservation areas deeply resented by local communities. This type of situation is very difficult to resolve.

Mining Conflicts

Irian Jaya is a land of rich mineral wealth, and mineral exploration and exploitation is increasing at an ever-alarming rate. Unfortunately, rich mineral deposits overlap frustratingly closely with virtually all the biologically valuable mountain reserves of the province. As world prices for mineral products climb, so do the threats to virtually every mountain range, and the search for oil and gas threatens the coastal plains as well as the slopes. There is no precedent of mining exploitation in Irian Jaya taking into account environmental protection, and the search for gold, silver, copper, coal, mica, nickel, uranium, and many other elements directly or indirectly threatens most if not all of the protected area system.

Agriculture

With agricultural development many lowland areas are increasingly converted to cash crop plantations, and both lowland and hill areas are subject to deforestation for food cultivation. A great many areas in the system suffer systematic encroachment, espe-

Table 4. Threatened marsupials and monotremes found, expected, or which may occur within the Protected Area System of Irian Jaya.

Lorentz National Park
Long-beaked echidna*, Stein's cuscus, great-tailed triok/striped possum, mouse bandicoot, Doria's tree-kangaroo, Weyland ringtail

Mamo-Foja National Park
Black-spotted cuscus, Clara or dimorphic bandicoot*, grizzled tree-kangaroo, Stein's cuscus

Jayawijaya Mountains Strict Nature Reserve
Long-beaked echidna, Clara or dimorphic bandicoot, mouse bandicoot, black-spotted cuscus, Stein's cuscus, great-tailed, triok/striped possum, Doria's tree-kangaroo, Goodfellow's tree-kangaroo

Weyland Mountains Strict Nature Reserve
Long-beaked echidna, Clara or dimorphic bandicoot, mouse bandicoot*, black-spotted cuscus*, Stein's cuscus*, great-tailed triok/striped possum*, Doria's tree-kangaroo, grizzled tree-kangaroo, white-throated tree-kangaroo, Weyland ringtail

Rouffaer River Wildlife Sanctuary
Clara or dimorphic bandicoot, black-spotted cuscus, grizzled tree-kangaroo

Wandamen Mountains Strict Nature Reserve
Long-beaked echidna, mouse bandicoot, black-spotted cuscus, Doria's tree-kangaroo*, grizzled tree-kangaroo, white-throated tree-kangaroo, Weyland ringtail

Cyclops Mountains Strict Nature Reserve
Clara or dimorphic bandicoot*, black-spotted cuscus*, grizzled tree-kangaroo*

Fakfak Mountains Strict Nature Reserve
Grizzled tree-kangaroo*, white-throated tree-kangaroo*

Kumawa Mountains Strict Nature Reserve
Grizzled tree-kangaroo*, white-throated tree-kangaroo

Arfak Mountains Strict Nature Reserve
Long-beaked echidna*, mouse bandicoot, Stein's cuscus, grizzled tree-kangaroo*, white-throated tree-kangaroo*, Arfak ringtail

Tamrau Utara Strict Nature Reserve
Long-beaked echidna, mouse bandicoot, Stein's cuscus*, grizzled tree-kangaroo, white-throated tree-kangaroo, Arfak ringtail

Tamrau Selatan Strict Nature Reserve
Long-beaked echidna*, mouse bandicoot, Stein's cuscus grizzled tree-kangaroo, white-throated tree-kangaroo, Arfak ringtail

Teluk Bintuni Strict Nature Reserve
Grizzled tree-kangaroo, white-throated tree-kangaroo

Wagura-Kote Zeno Forest Wildlife Sanctuary
No information available

Wasur Wildlife Sanctuary/Rawa Biru Strict Nature Reserve
Bronze quoll, red-cheeked dunnart, chestnut dunnart, Fly River bandicoot

Kumbe-Merauke Strict Nature Reserve
Bronze quoll, red-cheeked dunnart, chestnut dunnart, Fly River bandicoot

Danau Bian Wildlife Sanctuary
Bronze quoll, red-cheeked dunnart, chestnut dunnart , Fly River bandicoot

Pulau Kimaam Wildlife Sanctuary
Bronze quoll, red-cheeked dunnart, chestnut dunnart, Fly River bandicoot

Yapen Tengah Strict Nature Reserve
Clara or dimorphic bandicoot*, black-spotted cuscus, grizzled tree-kangaroo*

Biak Utara and Supiori Strict Nature Reserves
Clara or dimorphic bandicoot

Pulau Numfor Strict Nature Reserve
Clara or dimorphic bandicoot

Waigeo Barat Strict Nature Reserve
Waigeo spotted cuscus

Batanta Barat Strict Nature Reserve
Grizzled tree-kangaroo, white-throated tree-kangaroo

Salawati Utara Strict Nature Reserve
Long-beaked echidna*, grizzled tree-kangaroo, white-throated tree-kangaroo

Misool Selatan Strict Nature Reserve
Grizzled tree-kangaroo, white-throated tree-kangaroo

*=Confirmed records (from Petocz and de Frates 1983; additional information from G. George, pers. comm.)

cially in areas declared and marked for conservation without consultation with the local community.

Forestry Conflicts

Forestry concessions overlap more than 60% of the protected area system, and even formally recognised conservation areas are not safe from re-classification and forest exploitation. Increasingly larger areas are devoted to supplying plywood mills, pulp mills, and chipping operations. Vast areas of lowland forests and mangrove forests are cleared for exploitation, and only the steepest terrain is inaccessible to forestry development.

Trade in Wildlife

There is a flourishing legal and illegal trade in wildlife species from Irian Jaya, which includes marsupials and monotremes. Protected species such as cuscus (*Spilocuscus* spp., *Phalanger* spp.) and tree-kangaroos (*Dendrolagus* spp.) in particular are favoured pets. In Indonesia, taking, trading, etc. in protected species is prohibited, except with Ministerial permission. Permits are also needed to catch, transport, and export unprotected species. In Irian Jaya there is an almost total disregard for the law on protected species (Petocz 1984), and it seems likely that the same is true for unprotected species (Callister 1990). The brown dorcopsis wallaby (*Dorcopsis muelleri*) and the Celebes cuscus (*Strigocuscus celebensis*) are also protected by Indonesian law (Anon. 1982).

Hunting

Traditional hunting of marsupials and monotremes is widespread in Irian Jaya by hunter-gatherer peoples, but at this time is not considered unduly threatening. Systematic hunting using firearms and motorised vehicles, however, such as is reported in the Wasur wildlife sanctuary/proposed game park, is definitely threatening agile wallaby (*Macropus agilis*) populations.

The Protection Status of Marsupials and Monotremes in Indonesia

Table 5 lists the species that are protected by Indonesian law against capture, possession, and trade. The decrees under which the species are listed are in square brackets. The species listed however are "holdovers" from Dutch regulations. These are not based on true conservation status, nor on field studies.

Doria's tree-kangaroo, *Dendrolagus dorianus* (photo by T. Flannery).

Table 5. Marsupial and monotreme species protected by Indonesian law.

Tachyglossidae
[Surat Keputusan Menteri Pertanian No. 66/Kpts/Um/2/1973]

Long-beaked echidna, *Zaglossus bruijni*

Phalangeridae
[Surat Keputusan Menteri Pertanian No. 247/Kpts/Um/4/1979]

(all species of the genus *Phalanger*)

Ground cuscus, *Phalanger gymnotis*
Common spotted cuscus, *Spilocuscus maculatus*
Grey or common cuscus, *Phalanger orientalis*
Black-spotted cuscus, *Spilocuscus rufoniger*
Stein's cuscus, *Phalanger vestitus*
Waigeo Island spotted cuscus, *Spilocuscus papuensis*
Mountain cuscus, *Phalanger carmelitae*
Silky cuscus, *Phalanger sericeus*

Macropodidae
[Surat Keputusan Menteri Pertanian No. 421/Kpts/Um/8/1970 and No. 247/Kpts/Um/4/1979]

(all species of the genus *Dendrolagus*)

Doria's tree-kangaroo, *Dendrolagus dorianus*
Goodfellow's tree-kangaroo, *D. goodfellowi*
Grizzled tree-kangaroo, *D. inustus*
White-throated tree-kangaroo, *D. ursinus*

(all species of the genus *Thylogale*)

Brown or dusky pademelon, *Thylogale brunii*
Red-legged pademelon, *T. stigmatica*

Threatened Marsupials in other Areas of Indonesia

Threatened marsupials expected to be found in other areas and islands of Indonesia (and in reserves) are listed Table 6. As noted in the introductory paragraphs, many of the islands surrounding and to the west of Irian Jaya are far worse off environmentally than the mainland, in particular, the Maluku Islands, Timor, Wetar, and Sulawesi. Far more information needs to be collected to establish properly the status of island faunal populations. Sulawesi for example, has three species of cuscus (endemic to Sulawesi and the nearby Peleng Islands), and two endemic cuscuses can be found on the Halmahera Group of the Maluku Islands. Though none of these marsupials are known to be endangered, conversion of habitat is likely to threaten them (G. George pers. comm.).

Conclusions

The complexity of the problem of conserving an adequate representation of Irian Jaya's ecosystems is an issue matched only by its global biological significance. Irian Jaya holds what is probably the largest continuous Old World rain-forest, second only in importance to the New World's Amazon basin. The same physiography that has influenced its remarkable biological diversity has encouraged a vast cultural and linguistic diversity. As Irian Jaya undergoes development and rushes to catch up with the rest of Indonesia and the world, environmental concerns will have to be respected, if this development is to be sustainable.

There are already too many global examples of environmental mismanagement to the detriment of humankind and our planet. Irian Jaya has the potential, through a determined linkage of development and conservation, of being an example of wise national and regional sustainable development. Conservation is, after all, for people, and conservation must be seen as an available and desirable land-use option.

Current conservation management efforts in Irian Jaya are being implemented cooperatively between PHPA (Directorate General of Forest Protection and Nature Conservation) and WWF, through the WWF Irian Jaya Program. Cooperative projects in progress include the implementation of conservation management in the Teluk Cenderawasih marine conservation area, the Arfak Mountains Nature Reserve, and the Cyclops Mountains Nature Reserve (the latter two are important for marsupials, see Table 4).

Currently being considered as additions to the Irian Jaya program by WWF and PHPA are management planning and implementation projects in the proposed Lorentz National Park, and in the proposed Wasur National Park. In terms of marsupial and monotreme conservation, these two areas are extremely important (see Table 4).

WWF and PHPA jointly monitor development that impacts gazetted or proposed conservation areas, as well as monitoring wildlife trade. Though still far from achieving the objective of efficient management of the Irian Jaya Protected Area System, systematic progress is being made both in areas under management, and in national and local awareness of conservation issues. The conservation of marsupials and montremes depends upon the success of this cooperative regime. More specific recommendations for conserving marsupials and monotremes are given in Chapter 6.

Goodfellow's tree-kangaroo, *Dendrolagus goodfellowi* (photo by G. Steer).

Table 6. Threatened marsupials expected in Indonesian Protected Areas outside Irian Jaya (adapted from van der Zon 1979).

Maluku Province: Aru and Kai Islands
Sminthopsis virginiae, red-cheeked dunnart (Aru)
Phalanger gymnotis, ground cuscus (Aru)

Conservation Areas:
- Pulau Baun Wildlife Reserve. Forested island; may include representative Aru fauna.
- Kai Besar Strict Nature Reserve. Forested limestone hills. May contain all Kai species.
- Pulau Kobroor Strict Nature Reserve. Mangroves and swamp forests. May contain all Aru species.

Maluku Province: Southern Maluku
Rhynchomeles prattorum, Seram bandicoot (Seram)

Conservation Areas:
- Manusela National Park. Relatively undisturbed rain forest.

Maluku Province: Northern Maluku
Phalanger ornatus, Moluccan cuscus (Halmahera, Bacan, Morotai, Ternate)
Phalanger rothschildi, Obi Island cuscus (Obi)

Conservation Areas:
- Gunung Sibela Wildlife Reserve. Steep forested mountain slopes, Bacan Is. Possibility of *Phalanger ornatus*.

Sulawesi
Strigocuscus celebensis, little Celebes cuscus (Sulawesi, Sangihe)
Ailurops ursinus, large Celebes cuscus (Sulawesi, Peleng, Talaud, Togian Is.)
Phalanger pelengensis, Peleng Island cuscus (Peleng, Sula)

Conservation Areas: (Data on Sulawesi is being collated)
- Pulau Seho Wildlife Reserve. Forest dominated by *Pometia tomentosa*. Possibility of Peleng Island cuscus.
- Pulau Taliabu Strict Nature Reserve. High steep forests, some largely logged lowland forest. Possibility of Peleng Island cuscus.
- Pulau Lifamatola Wildlife Reserve. Easternmost of the Sula Islands. Possibility of little Celebes cuscus.
- Pulau Obi Wildlife Reserve. Forests on steep terrain. Possibility of Obi Island cuscus.

Timor, Wetar, Leti
Conservation Areas: There are no conservation areas on Wetar and Leti. (Data on Timor is being collated).

3. Conservation of Marsupials and Monotremes in Papua New Guinea

Lester Seri[1]

The often stated assertion that Papua New Guinea is in a fortunate situation with large portions of its natural environment still intact is no longer appropriate given its current stage of development. Socio-economic and environmental changes have come about very rapidly since Papua New Guinea attained independence fifteen years ago. The country's human population of about 4 million, growing at 2.7 percent annually, will reach 6 million or more by the turn of the century (Patchett 1991). Multi-million kina development projects involving logging, agriculture, mining, etc. have increased substantially, and will no doubt continue to accelerate. Logging operations in the country have more than tripled in area in less than a decade. It is predicted that by the turn of the century, assuming the same rate of timber extraction as at present, much of the country's easily accessible forested areas will be cleared.

The resulting impact on the natural environment will no doubt be substantial, particularly on the native forests which provide the habitats for the major component of the country's terrestrial fauna. The principal threat to the continued existence of the fauna is, therefore, loss of habitat. Any hope of safeguarding the continued existence of the fauna will depend largely upon protection of these critical habitats. Habitat conservation must be the first and foremost priority in any plan of action for the conservation of marsupials and monotremes in Papua New Guinea.

Woodlark Island cuscus, *Phalanger lullulae* (photo by T. Flannery).

Marsupial and Monotreme Species of Papua New Guinea

The Island of New Guinea has a diverse mammal fauna similar to that of Australia, including three sub-classes: monotremes, marsupials, and eutherians, all occupying a wide variety of habitats ranging from the coast to tropical rainforest and alpine grassland. There are nearly 200 non-marine mammal species on the island, the major components of which are forest-dwelling. About 78 of the estimated 200 mammals are marsupial-monotreme species, including those that occur on nearby Indonesian islands.

On the basis of George (Appendix 1), Papua New Guinea alone accounts for 64 of the 78 marsupial and monotreme species with about 18 endemics. Irian Jaya accounts for 58 of the 78 species, with 13 endemics. Eleven of the 18 Papua New Guinea endemics occur on off-shore islands, with five exclusively insular. Table 7 lists the threatened marsupials on Papua New Guinea's off-shore islands.

Most of the species are confined to the mainland, particularly within the biogeographic regions of the central cordillera, the north-south complex mosaic of mountain ranges, lowlands, and the savannah plains and the south and south east regions. Populations of geographically different forms also occur in these areas. The species are generally distributed along the altitudinal gradient with several overlapping in their distributional ranges. The off-shore islands are generally depauperate in species composition, with only a few endemics, reflecting their past land connections and isolation from the species-rich mainland.

Although considerable interest has been shown in the mammalian fauna of Papua New Guinea for over a century, a great deal remains unknown. Study of the fauna is still in its exploratory phase, while the current taxonomy is very much in flux. Informa-

[1]Department of Environment and Conservation, Port Moresby, Papua New Guinea.

tion on the biology, ecology, population, distribution, and the extent of the utilization of the fauna has yet to be properly documented.

With such limitations in our present knowledge about the fauna it is quite difficult to effectively attempt classification of the known species into the IUCN categories of Endangered, Vulnerable, or Rare. There is no doubt, however, that a number of species are threatened, but to what degree is difficult to say with any certainty.

The current IUCN Red List of Threatened Animals (1990) lists one monotreme and eight marsupial species from Papua New Guinea as either Endangered, Vulnerable, or Rare based largely on George (1979). The list is made up of the larger species which are subject to considerable hunting pressure, and disregards many of the infrequently encountered smaller species.

At the time the IUCN list was prepared, the broadbent bandicoot (*Peroryctes broadbenti*) was considered to be much more common because of confusion in the literature with the Raffray bandicoot (*P. raffrayanus*) (George & Maynes 1990). Only three specimens of this large bandicoot are currently known to science.

Six new species have been described since 1979 (the bronze quoll (*Dasyurus spartacus*), Torricelli glider (*Petaurus abidi*), Telefomin cuscus (*Phalanger matanin*), Kiriwina bandicoot (*Echymipera davidi*), Fly River bandicoot (*Echymipera echinista*), and Scott's tree-kangaroo (*Dendrolagus scottae*). Most have very small known populations and may be endangered, however, the bronze quoll has recently been found to be very common in the Morehead-Mibini area of the Western Province of Papua New Guinea (P. Woolley pers. comm.). The lowland tree-kangaroo (*Dendrolagus spadix*) has also been recognised as a distinct species by Flannery (1990), who lists it as "rare".

Many of the infrequently encountered small species need special attention, if only to clarify their conservation status. Flannery (1990) lists the speckled dasyure (*Neophascogale lorentzii*), short-furred dasyure (*Murexia longicaudata*), broad-striped dasyure (*Murexia rothschildi*), and Papuan planigale (*Planigale novaeguineae*) as "uncommon"; the three-striped dasyure (*Myoictis melas*) and Papuan bandicoot (*Microperoryctes papuensis*) as "rare". In addition to these uncommon and rare species, other species that occur in areas of high human population should be considered "Potentially Vulnerable" as environmental changes rapidly accelerate. Flannery also suggests that all mammals over 6 kg in weight are in trouble wherever human population densities are high. All reproduce slowly and are easy targets for hunting. Flannery (pers. comm.) suspects that all tree-kangaroos, echidnas, and many wallaby species are now locally extinct in much of their previous range.

This Action Plan has identified three endangered, eight vulnerable and twelve potentially vulnerable marsupial and monotreme species in Papua New Guinea (see Table 8). Recommendations for revised IUCN Red Data Book listings are given in Chapter 6, Table 23. A complete listing indicating the conservation status assigned to all Papua New Guinea marsupials and monotremes by the Action Plan is given in Chapter 6, Table 21. Brief details of reasons for listing, distribution, habitat type and recommended conservation action for endangered, vulnerable, and potentially vulnerable species are given in Chapter 6, Tables 22 and 24 respectively.

Juvenile Scott's tree-kangaroo, *Dendrolagus scottae* (photo by T. Flannery).

Table 7. Threatened marsupials on off-shore islands of Papua New Guinea.

Species	Island
Matschie's tree-kangaroo *Dendrolagus matschiei*	Umboi
Black dorcopsis *Dorcopsis atrata*	Goodenough*
Woodlark Island cuscus *Phalanger lullulae*	Woodlark*/Alcester*
Kiriwina bandicoot *Echymipera davidi*	Kiriwina*
Fergusson Island striped possum *Dactylopsila tatei*	Fergusson*

Papua New Guinea has about 800 or more islands and archipelagoes on which mammals (both placental and marsupial groups) occur. Information on the presence of the two mammal groups is substantially lacking, as little detailed surveying of the islands has been carried out to date. The information above is based on known recorded data. The great majority of the islands remain to be comprehensively surveyed.

*Species endemic to the island

Conservation in Papua New Guinea

The World Bank-sponsored Tropical Forest Action Plan (TFAP) for Papua New Guinea was presented to the government in early 1990. It was a plan aimed at the integrated development of forestry resources, but on a sustainable basis. However, the Papua

Table 8. Threatened Papua New Guinea marsupials and monotremes.

Common Name	Scientific Name	Status
Long-beaked echidna	*Zaglossus bruijni*	Vulnerable
Bronze quoll	*Dasyurus spartacus*	Potentially Vulnerable
Chestnut dunnart	*Sminthopsis archeri*	Potentially Vulnerable
Red-cheecked dunnart	*Sminthopsis virginiae*	Potentially Vulnerable
Kiriwina bandicoot	*Echymipera davidi*	Potentially Vulnerable
Clara or dimorphic bandicoot	*Echymipera clara*	Potentially Vulnerable
Fly River bandicoot	*Echymipera echinista*	Potentially Vulnerable
Broadbent/Giant bandicoot	*Peroryctes broadbenti*	Endangered
Woodlark Island cuscus	*Phalanger lullulae*	Endangered
Telefomin cuscus	*Phalanger matanin*	Vulnerable
Stein's cuscus	*Phalanger vestitus*	Potentially Vulnerable
Black-spotted cuscus	*Spilocuscus rufoniger*	Vulnerable
Doria's tree-kangaroo	*Dendrolagus dorianus*	Vulnerable
Goodfellow's tree-kangaroo	*Dendrolagus goodfellowi*	Vulnerable
Matschie's tree-kangaroo	*Dendrolagus matschiei*	Potentially Vulnerable
Lowland tree-kangaroo	*Dendrolagus spadix*	Potentially Vulnerable
Scott's tree-kangaroo	*Dendrolagus scottae*	Endangered
Grizzled tree-kangaroo	*Dendrolagus inustus*	Potentially Vulnerable
Black dorcopsis	*Dorcopsis atrata*	Potentially Vulnerable
Macleay's dorcopsis	*Dorcopsulus macleayi*	Vulnerable
Great-tailed triok/Striped possum	*Dactylopsila megalura*	Vulnerable
Fergusson Island Striped possum	*Dactylopsila tatei*	Potentially Vulnerable
Torricelli/Northern glider	*Petaurus abidi*	Vulnerable

New Guinea Department of Environment and Conservation (DEC) was concerned by what it saw as serious problems in the conservation and environmental management aspects of the report (Patchett 1991).

While generally endorsing the World Bank's report in preparation for a TFAP Round Table Conference in 1990, the DEC and other relevant government departments developed their own conservation program (some 15 projects in all) that now constitute Papua New Guinea's "National Forest and Conservation Action Plan" (NFCAP). The broad program components are geared towards truly comprehensive environmental management regimes, which will sustainably utilise natural forest resources while at the same time recognizing the national requirement to protect Papua New Guinea's immense biological diversity.

The significance of the NFAP proposals for the conservation of marsupials and monotremes in Papua New Guinea is enormous. As in the case of Irian Jaya, the only long-term and practical means of conserving a significant proportion of the existing mammalian fauna is via the establishment of a comprehensive system of protected areas, though stressing that this can only be done with the full support of local landowners. While there must be full provision for the protection of marsupials and monotremes within each protected area management plan, the concept of "single species recovery plans", applicable for instance in Australia, simply cannot be carried out the same way in Papua New Guinea.

Therefore, the following comments on current conservation mechanisms in Papua New Guinea, are given on the understanding that such mechanisms may be substantially changed in the near future, and that they will also form the core opportunity for implementing successful wildlife conservation programs.

Legislation

(Adapted from Eaton 1986)

The Constitution of Papua New Guinea specifically allows for protection and wise use of the country's natural resources and environment: "We declare our fourth goal to be for Papua New Guinea's natural resources to be conserved and used for the collective benefit of us all, and be replenished for the benefit of future generations.

We accordingly call for:

- wise use to be made for our natural resources and the environment in and on the land or seabed, in the sea, under the land, and in the air in the interest of our development and in trust for future generations.
- the conservation and replenishment, for the benefit of ourselves and posterity, of the environment and sacred, scenic, and historical qualities; and
- all necessary steps to be taken to give adequate protection to all our valued birds, animals, fish, insects, plants, and trees."

At present, there are nine primary pieces of legislation which protect the natural resources and environmental in Papua New Guinea:

- Fauna (Protection and Control) Act 1974
- Environment Contaminants Act (1978)
- Conservation Areas Act (1978)

- National Parks Act (1982)
- Environment Planning Act (1978)
- International Trade of Endangered Species of Fauna and Flora Act (1979)
- Crocodile Trade Protection and Control Act (1984)
- Water Resources Act (1982)
- Customs Regulations (1973)

Current Wildlife Protection and Protected Areas Mechanisms

The taking and possession of protected and restricted species is controlled under the Fauna (Protection and Control) Act, 1974 and is subject to licensing (see Tables 9 and 10). Taking and domestic trade in protected wildlife is not permitted, except that Automatic Citizens of Papua New Guinea may take wildlife for traditional purposes. Although there are no controls over the taking and domestic trade in unprotected wildlife, the consent of landholders is required (Callister, 1990).

The commercial export of live species from Papua New Guinea is prohibited, and exports can only occur for scientific and educational purposes. The export of vertebrate species can only be permitted to approved zoological or scientific institutions (Callister 1990). A number of protected and restricted species can also be found in the Appendices to CITES, and a full list of Papua New Guinea species recognised by CITES is given in Chapter 6, Table 23.

Table 9. Protected marsupial and monotreme species in Papua New Guinea (conditions apply to exports only).

Tachyglossidae
Long-beaked echidna, *Zaglossus bruijni*—CITES App. II*
Fauna (Protection and Control) Act 1974
International Trade (Fauna and Flora) Act 1979

Macropodidae
Doria's tree-kangaroo, *Dendrolagus dorianus*
Fauna (Protection and Control) Act 1974

Goodfellow's tree-kangaroo, *D. goodfellowi*
Fauna (Protection and Control) Act 1974

Grizzled tree-kangaroo, *D. inustus*—CITES App. II*
Fauna (Protection and Control) Act 1974
International Trade (Fauna and Flora) Act 1979

Matschie's tree-kangaroo, *D. matschiei*
Fauna (Protection and Control) Act 1974

Lowland tree-kangaroo, *D. spadix*
Fauna (Protection and Control) Act 1974

Black dorcopsis, *Dorcopsis atrata*
Fauna (Protection and Control) Act 1974

Phalangeridae
Black-spotted cuscus, *Spilocuscus rufoniger*
Fauna (Protection and Control) Act 1974

*App. II=species in which international commercial trade is only authorised under permit.

Table 10. Marsupial species in Papua New Guinea for which restrictions are placed on their use.

Species subject to restrictions and listed in CITES Appendices:

Macropodidae
Grizzled tree-kangaroo, *Dendrolagus inustus*—App. II

Species subject to restrictions not listed in CITES:

Tachyglossidae
Short-beaked echidna, *Tachyglossus aculeatus*

Macropodidae
Black dorcopsis, *Dorcopsis atrata*
Macleay's dorcopsis, *D. macleayi*

Phalangeridae
Feather-tailed possum, *Distoechurus pennatus*
Black-spotted cuscus, *Spilocuscus rufoniger*

Dasyuridae
New Guinea planigale, *Planigale novaeguineae*

Peroryctidae
Clara bandicoot, *Echymipera clara*
Giant bandicoot, *Peroryctes broadbenti*

*App. II=species in which international commercial trade is only authorised under permit.

The three most important pieces of legislation for the protection and conservation of wildlife and establishment of protected areas in Papua New Guinea are the National Parks, Conservation Areas, and Fauna (Protection and Control) Acts, although the Environment Planning Act can also be effective in protecting species and habitats.

The conservation provisions under the National Parks Act are similar in kind to those employed in establishing national parks in many developed countries, based largely on the principle of alienation of land from landowners. Provisions are also made for leasing and accepting of land. Given that in Papua New Guinea a large proportion (98%) of the land is owned by indigenous peoples (Patchett 1991), areas so far declared as National Parks under the Act are on government owned land.

The major objective of the National Parks Act is "the conservation of sites and areas having particular biological, topographical, geological, historical, scientific, or social importance."

There are currently three National Parks declared and gazetted. There are a number of National Parks awaiting final gazettal and many other proposed areas awaiting the necessary settlement of land acquisition and agreement with the landowners. The Act's intent suffers from lack of implementation associated with land tenure problems.

The Conservation Areas Act enables conservation areas to be established on land held under customary tenure. There are provisions for a management plan for each area and there can be no changes in land use or development in the area which are not in accordance with this plan or do not have the written permission of the Minister for Environment and Conservation. Unfortunately, this legislation is yet to be implemented and at present no conservation areas have been declared under the statute. The Act allows for establishment of a National Conservation Council which would implement the Act through the Department of Environment and Conservation.

The Fauna (Protection and Control) Act dates from 1966. It contains provisions for the protection of certain wildlife species that are considered to be endangered. Part III of this Act states that protected species may be hunted only by indigenous peoples using traditional hunting methods for customary non-commercial purposes. In addition, the Act provides for the establishment of three types of conservation areas which may be developed on customary land. These are Wildlife Sanctuaries (WS), Species Protected Areas (SPA), and Wildlife Management Areas (WMA). In Wildlife Sanctuaries, all fauna are protected unless otherwise specified by the Minister for Environment and Conservation. The reverse applies in Species Protected Areas where it is prohibited to take and kill animals specified in the gazettal of the area. Wildlife Management Areas may be declared after consultation with the landowners and local government authorities. A wildlife management committee is formed from representatives of the landowners and rules are made for the protection and management of wildlife in the area. The rules may also provide for licences, charges, and the collection of royalties. The procedures for the development of Wildlife Management Areas are shown in Table 11.

Table 11. Procedure for the development of a Wildlife Management Area in Papua New Guinea.

1. Customary land-owners inform government of wildlife problems.
2. Discussions between land-owners and government field officers.
3. Final meeting to decide boundaries, committee members, and regulations.
4. Report by field officer to Wildlife Division.
5. Department of Lands prepare legal boundaries description.
6. Legislative Counsel prepares declaration which includes boundaries, committee members, and the name of the area.
7. Declaration to Government Printer for printing in National Gazette.
8. Statutory Instrument signed by Minister for Environment and Conservation.
9. Instrument submitted to National Executive Council for information.
10. Governor-General approves Statutory Instrument.
11. Publication of rules in National Gazette.

There are, at present, 17 Wildlife Management Areas, five Wildlife Sanctuaries, and two Species Protected Areas. In addition, the Papua New Guinea protected areas system contains two Provincial Parks; two Nature Reserves; one Historic Park; and one Historic Reserve. These cover an area equal to 2.3% of the land mass of Papua New Guinea (G. Kula pers. comm.). All declared sites are listed in Table 12 and their location is shown in Figure 2. There are proposals for a further 83 areas (mostly WMAs) currently with the Papua New Guinea government for consideration. There is currently no species checklist for Papua New Guinea's protected areas system.

Other Forms of Wildlife Protection

"Green belts" or "buffer zones" of up to 100 m (50 m either side of permanent water courses) are required in major forestry operations, as environmental safeguards and to provide wildlife corridors or refuge areas for forest dwelling fauna. This practice is allowed for under the Environmental Planning Act (1978) where Environmental Plans are perquisite to government approval of development projects. The effectiveness of such practices in protecting wildlife has yet to be determined.

Problems in Implementing The Protected Areas System

There are a number of matters that can inhibit effective implementation of the conservation or protected areas systems in Papua New Guinea. These include:

1. A lack of staff and inadequate funds make effective implementation of conservation areas, and monitoring and management of declared protected areas very difficult. At present, a long list of proposed conservation areas remains to be acted upon, due to both land ownership conflicts and lack of resources.
2. The possibility of cash income from development of resources in individual or group owned land makes landowners reluctant to set aside portions or all of their land for conservation purposes. There are situations of "change of heart" among landowners of declared conservation areas wanting to apportion part of a declared area for forestry development as a direct result of development activities going on adjacent to the reserve area.
3. A lack of effective environment education or awareness campaigns, and a lack of technical training of government and community groups.

This Action Plan recognises, however, that current and proposed protected areas in Papua New Guinea cannot be solely set aside from the question of other resource use. As noted by the WWF Papua New Guinea Protected Areas Rehabilitation Program, 1991: "The patterns of land ownership and human habitation throughout the country make it essential for nature protection to

Table 12. Declared Protected Areas in Papua New Guinea, 1991.

Category of Protected Area	No.	Name	Area (ha)	Province
National Park	3	Varirata National Park	740	Central
		Jimi Valley National Park	4,180	Western Highlands
		McAdam National Park	2,080	Morobe
Nature Reserve	2	Talele Island Nature Reserve	1	East New Britain
		Nanuk Nature Reserve	4	East New Britain
Species Protected Area	2	Baniara Island SPA	100	Milne Bay
		Lihira SPA	—	New Ireland
Provincial Park	2	Mt. Gahavisuka Provincial Park	77	Eastern Highlands
		Lorko Provincial Park	740	West New Britain
Wildlife Sanctuary	5	Baiyer River Wildlife Sanctuary	102	Western Highlands
		Balek Wildlife Sanctuary	470	Madang
		Randa Wildlife Sanctuary	41,922	Madang
		Moitaka Wildlife Sanctuary	—	Central
		Oia Mada Wa Wildlife Sanctuary	22,840	Milne Bay
Wildlife Management Area	17	Iomare WMA	3,827	Central
		Zo-oinga WMA	1,510	Central
		Mojirau WMA	5,079	East Sepik
		Aird Hills (Meiru) WMA	3,984	Gulf
		Bagiai-Karkar Island WMA	13,760	Madang
		Ndrolowa WMA	5,850	Manus
		Lake Lavu WMA	2,640	Milne Bay
		Sawataetae WMA	700	Milne Bay
		Mt. Kaindi WMA	150,835	Morobe
		Mt. Nuserang WMA	22	Morobe
		Pirung (Eight Islands) WMA	43,200	North Solomons
		Crown Island WMA	58,969	Madang
		Siwi Utami WMA	12,540	Southern Highlands
		Maza WMA	184,230	Western Highlands
		Tonda WMA	590,000	Western Highlands
		Garu WMA	8,700	West New Britain
		Pokili WMA	9,840	West New Britain
Historic Reserve	1	Namanatabu Historic Reserve	27	Central
Historic Park	1	Cape Wom Historic Park	105	Eastern Highlands

be integrated with other uses. In some protected areas, conservation values must be sustained alongside a variety of other use and development activities. Multiple-use Protected Areas such as Wildlife Management Areas are particularly important in this regard. This requirement will be a major consideration in the proposed rehabilitation and expansion program."

Reserve or Protected Area Systems Proposals that will Benefit Marsupial and Monotreme Conservation

Detailed proposals have been made in the past for a comprehensive national reserve system for Papua New Guinea (Diamond undated; Menzies 1976; Schodde 1973; Schultze-Westrum 1973). The World Bank (1990) Tropical Forest Action Plan also suggested a comprehensive protected areas system involving the dedication of four World Heritage Sites implemented as part of the NFCAP program.

Diamond's (undated) proposed "Natural Reserve System for Papua New Guinea" was based on his comprehensive ecological studies of avifauna over several years. He recommended a total of 22 reserves and provided indicative area sizes.

In highlighting the country's problems and inadequacies in effective implementation of conservation programs, Menzies (1976) suggested smaller reserve areas, as they would be easier to establish. He also noted that the fauna of Papua New Guinea consists largely of small animals, perhaps making small reserves more viable.

Northern glider, *Petaurus abidi* (photo by T. Flannery).

secure incentive agreements, be made available to landowners. Mechanisms for developing and implementing this objective form an important part of the DEC Strategic Plan (the first draft of which was prepared in December 1991), and are the basis for the Papua New Guinea Global Environment Facility (GEF) Project.

Integrated Conservation and Development

The participants of the Global Environment Facility endorsed, and have committed funds for, an Integrated Conservation and Development (ICAD) Program in Papua New Guinea. This initiative was a consequence of the series of linked initiatives that will build upon several first-stage projects currently in the implementation stage, including:

DEC Strategic Plan will establish clear priorities for a range of organizational objectives and provide, in a single document, a cohesive planning framework for the next five years in definitive terms, and for ten years, in a more general context. The plan will also provide an investment analysis of organizational strengthening needs. (Co-financed by World Bank/UNDP, and New Zealand government.)

Conservation Needs Assessment (CNA), which is assembling and analyzing the existing information on Papua New Guinea's biodiversity, demographics, land tenure and land-use systems, ongoing research programs, maps and other geographic information as well as information on non-extractive forest products and their potential markets. (Funding provided by USAID through WWF's Biodiversity Support Program.)

Protected Areas Rehabilitation Program (Stage I) comprises a review of Papua New Guinea's existing protected areas, including legislation, management, enforcement, policy and conservation value, as well as existing levels of awareness, support and involvement of NGOs and local communities. The review will design and lay the foundations for the subsequent comprehensive program of rehabilitating and expanding selected categories of conservation areas reserves and parks. (Funding provided by the European Community and execution by WWF.)

Rapid Resource Appraisal Project, which will utilise Papua New Guinea Resource Information System (PNGRIS), CSIRO (Commonwealth Scientific and Industrial Research Organisation), and other data to evaluate the extent and condition of the forest resource in Papua New Guinea, including commercial, traditional timber, and non-timber uses. (Funded by the Australian International Development Assistance Bureau—AIDAB.)

Representative Protected Areas System Design has as a primary objective a manual that will provide guidance for implementation and application of criteria for the identification, classification, justification, and establishment of protected areas in Papua New Guinea.

Conclusions

The Action Plan recommendations (see chapter 6) are made in full awareness of the Papua New Guinea Government's National Forest and Conservation Action Plan (NFCAP) and represent in particular the desire to implement "Conservation Areas" in cooperative projects with landowners. The NFCAP scheme proposes to establish agreed environmental management and incentive packages with landowners for each of the Conservation Areas with funding to come primarily from international donors, both governmental and non-governmental and the progressive implementation of the proposed Environment and Conservation Trust Fund. This concept is another element of the GEF project that will be fully investigated as part of the effort necessary to ensure that conservation programs in Papua New Guinea have a high level of internal sustainability.

The achievement of conservation and ecologically sustainable development in Papua New Guinea depends upon an unprecedented degree of cooperation and collaboration between various agencies of Papua New Guinea, together with the national and international NGO community with established credentials and a genuine interest in the well-being of the Papua New Guinean people and the conservation of the nation's natural resource base.

The reserve system proposal, whatever shape it may finally take, must ensure that it adequately caters to the biological needs of Papua New Guinea's unique marsupial and monotreme species. In combination with proposed upgraded legislative responsibilities to conserve wildlife, and long-term well-directed wildlife education programs, the prospects for maintaining Papua New Guinea's endemic marsupials and monotremes in all their diversity are encouraging.

Distribution of Endangered and Vulnerable Marsupials and Monotremes in Indonesia and Papua New Guinea

The information on the distribution of Endangered and Vulnerable marsupials and monotremes in Indonesia (Irian Jaya) and Papua New Guinea is reproduced from Flannery (1990).

Flannery notes that the "...distribution maps... are based on a comprehensive literature survey, specimens held in the Australia Museum and National Wildlife Collections, and limited material that I examined in North American institutions during 1983. Type localities, where on mainland New Guinea, are marked with a red (black) star, although I have been unable to mark some type localities as they are too vague (eg. "southeastern Papua"). Fossil and subfossil occurrences are marked with a black cross, while occurrences based on the collection of a modern animal are marked with a black dot.

I have been conservative in constructing these maps in that if doubt exists as to the identity of a specimen cited in the literature, or a locality is doubtful, then the record has been excluded."

The distribution maps for the Seram bandicoot (*Rhynchomeles prattorum*), the Woodlark Island cuscus (*Phalanger lullulae*), and Scott's tree-kangaroo (*Dendrolagus scottae*) developed for this Action Plan, show the island location for the first two species with a black arrow, and the approximate mainland location for the latter species. Amendments (G. George, pers. comm.) have been made to Flannery's maps for the Giant bandicoot (*Peroryctes broadbenti*) and the black-spotted cuscus (*Spilocuscus rufoniger*).

Table 13. Recommended natural reserves for Papua New Guinea (Diamond, undated).

This list consists of 22 reserves designed to incorporate most of the genetic wealth of Papua New Guinea. For each reserve a minimum area is recommended that will keep extinction rates low for the reserve's species. Except for the New Ireland Reserve (number 7) and possibly the D'Entrecasteaux Reserve (number 9), each reserve should be continuous, and should not consist of separate pieces isolated from each other. Where reserves are proposed to include both lowlands and highlands (reserves 1 through 9), this is because different sets of species occur in the lowlands and highlands. Thus, it is necessary to keep both the lowland and highland parts of such reserves, and not to place all the area in the highland just because of fewer people and less convenient access for logging.

1. **Fly-to-Sepik Reserve.** Lowlands of Western District in the drainage of the Fly and Strickland Rivers (including some extensive savannahs and marshes as well as rainforest) north across the Central Range both east and west of the Strickland Gorge, and including the highest mountains of this region; into the lowlands of the upper Sepik River. If possible, extending continuously north to the Bewani-Torricelli reserve. At least 5,200,000 hectares.

2. **Southeastern Reserve.** From the lowlands of Central Province somewhere between Bereina and Samarai, across the Owen Stanley Range, down into the lowlands of Morobe and Northern Provinces somewhere between Lae and Milne Bay. The location may depend on practical considerations but should contain some high peaks with alpine grassland, such as Mt. Albert-Edward, Mt. Victoria, and Mt. Suckling. About 2,600,000 hectares.

3. **Huon Reserve.** Mountains of the Huon Peninsula from an altitude of 450 meters up to the highest peaks of the Finisterre and Saruwaged Ranges, and preferably continuing down to low elevations in the foothills of the Markham Valley and north coast. 780,000 hectares.

4. **Adelbert Reserve.** Adelbert Mountains, including the highest peaks and extending down to the foothills. 78,000 hectares.

5. **Bewani-Torricelli Reserve.** Bewani and Torricelli mountains, between Dreikikir and the Irian Jaya border, above an altitude of 450 meters, up to the highest summits. If possible, should connect with the Fly-to-Sepik Reserve. 260,000 hectares.

6. **New Britain Reserve.** This should be centred on one of the high mountain ranges: either the Whiteman Range, Nakanai Mountains, or mountains of the Gazelle Peninsula. The choice among these three ranges depends on practical considerations. From the highest summits down to the lowlands of both the northern and southern watersheds. 1,300,000 hectares.

7. **New Ireland Reserve.** The most practical location would be the sparsely populated southeast end, from the lowlands up to the highest summits. If so, there should be a separate reserve at the Kavieng end or else at New Hanover, where populations show some genetic differences from the southeast end. 260,000 hectares.

8. **Bougainville Reserve.** Lowlands to some of the highest peaks. Mt. Balbi might be a good focus because of its scenic and volcanic interest. 390,000 hectares.

9. **D'Entrecasteaux Reserve.** Lowlands of either Fergusson, Goodenough, or Normanby Island, plus mountains of Goodenough. Total of 104,000 hectares.

10. **Manus Island.** 78,000 hectares.

11. **Mussau Island.** 23,400 hectares.

12. **Kiriwina Island.** 7,800 hectares.

13. **Woodlark Island.** 39,000 hectares.

14. **Misima Island.** 7,800 hectares.

15. **Tagula Island.** 39,000 hectares.

16. **Rossel Island.** 13,000 hectares.

17. **Dyaul Island.** 5,200 hectares.

18. **Tabar Islands** (Tabar or Tatau). 10,400 hectares.

19. **Lihir Islands** (Lihir). 7,800 hectares.

20. **Tanga Islands** (Tanga or Boang). 5,200 hectares.

21. **Feni Islands** (Feni). 5,200 hectares.

22. **Nissan Islands.** 1,300 hectares.

Table 14. Recommendations for areas in which large reserves might be placed (Schodde, 1973).

Sector	Location	Altitudinal Range (Meters)	Major Vegetation Habitats
1	Southern scarp of Arfak Mountains, Vogelkop, Irian Jaya	0-3,000	Mangrove swampland; forested swampland; lowland and hill rainforest; montane rainforest; subalpine meadowland
2	Southern scarp of Nassau Range, Irian Jaya	0-5,000	Mangrove swampland; forested swampland; lowland and hill rainforest; montane rainforest; alpine and subalpine meadowland
3	Northern scarp of Van Rees Mountains-lower Mamberamo River, Irian Jaya	0-2,000	Forested swampland; lowland and hill rainforest; montane rainforest; some anthropogenous grassland
4	Meervlakte and northern scarp of Oranje Range, Irian Jaya	50-4,300	Forested swampland; lowland and hill rainforest; montane rainforest; alpine and subalpine meadowland
5	Eastern Star Mountains, Irian Jaya and Papua New Guinea	100-4,600	Lowland and hill rainforest; montane rainforest; alpine and subalpine meadowland; some anthropogenous grassland; *Baeckea*-ericaceous heathland
6	Lacustrine region, middle Fly River, Irian Jaya and Papua New Guinea	20-100	Grassed swampland; some forested swampland; lowland rainforest
7	Trans-Fly plains, Irian Jaya and Papua New Guinea	0-30	Forested swampland; gallery rainforest; monsoon forest; eucalypt-paperbark woodland; *Sinoga* heathland; some mangrove swampland
8	Bewani Mountains and northern Sepik watershed, Papua New Guinea	30-2,000	Forested swampland; lowland and hill rainforest; montane rainforest; some anthropogenous grassland
9	Headwaters of Strickland River-Hunstein Range, Papua New Guinea	50-3,100	Lowland and hill rainforest; montane rainforest; some anthropogenous grassland.
10	Mt. Bosavi-Mt. Kerewa-Hegigo River area, Papua New Guinea	50-3,700	Lowland and hill rainforest; montane rainforest; alpine and subalpine grassland; some forested swampland; some anthropogenous grassland
11	Northern scarp of Bismarck Range, Papua New Guinea	50-4,500	Lowland and hill rainforest; montane rainforest; alpine and subalpine grassland; some anthropogenous grassland
12	Southern scarp of Kratke Range-Purari-Kikori Rivers, Papua New Guinea	0-3,400	Mangrove swampland; forested swampland; lowland and hill rainforest; montane rainforest; some anthropogenous grassland
13	Southern Adelbert Range, Papua New Guinea	50-1,900	Lowland and hill rainforest; montane rainforest
14	Scarps of Saruwaged and Cromwell Ranges, Papua New Guinea	10-4,100	Lowland and hill rainforest; montane rainforest; alpine and subalpine meadowland; anthropogenous grassland
15	Vanapa-Laloki-Kemp Welch River systems and adjacent ranges (Astrolable Range, Mt. Maguli), Papua New Guinea	0-1,800	Lowland hill rainforest; eucalypt-paperbark woodland; monsoon forest; gallery rainforest; some grassed and forested swampland; some mangrove swampland
16	Scarps of Mt. Suckling massif and Keveri Hills, Owen Stanley Range, Papua New Guinea	20-3,700	Lowland and hill rainforest; alpine and subalpine grassland; eucalypt woodland; some anthropogenous grassland
17	Nakanai Mountains and their northern watershed, New Britain	0-2,600	Forested swampland; lowland and hill rainforest; montane rainforest
18	Hans Meyer Range and adjacent lowlands, New Ireland	0-2,700	Lowland and hill rainforest; montane rainforest

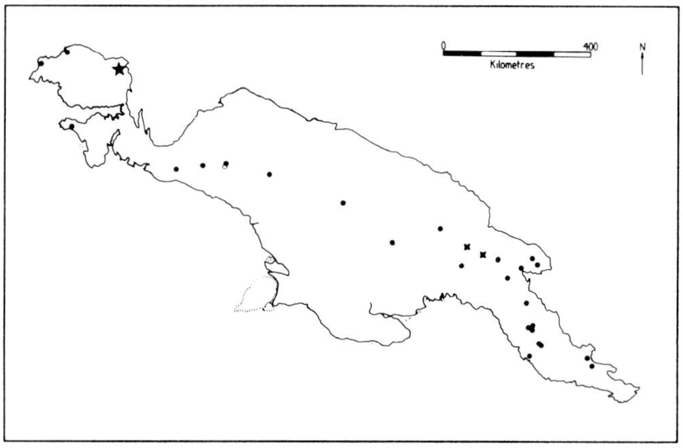

Figure 4. Distribution of the long-beaked echidna, *Zaglossus bruijni.*

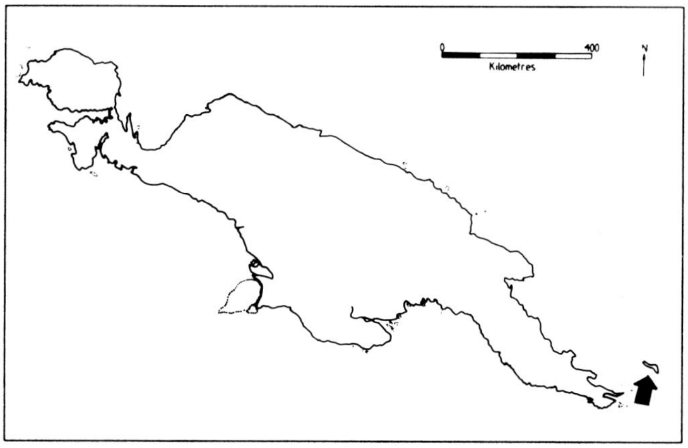

Figure 7. Distribution of the Woodlark Island cuscus, *Phalanger lullulae.*

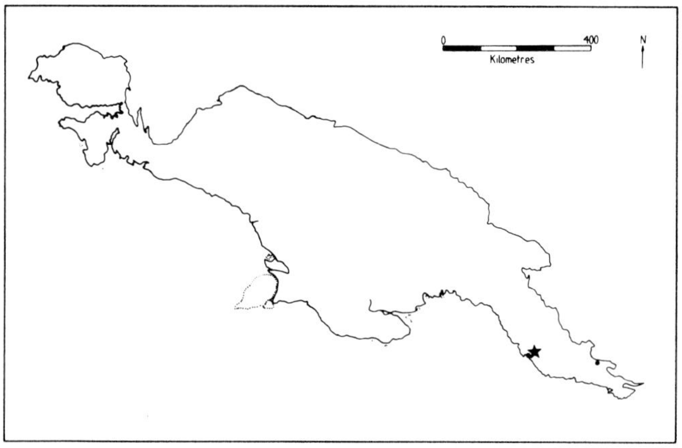

Figure 5. Distribution of the giant bandicoot, *Peroryctes broadbenti.*

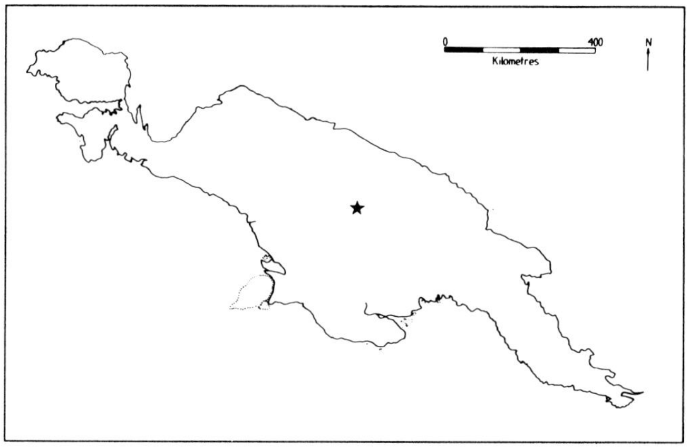

Figure 8. Distribution of the Telefomin cuscus, *Phalanger matanim.*

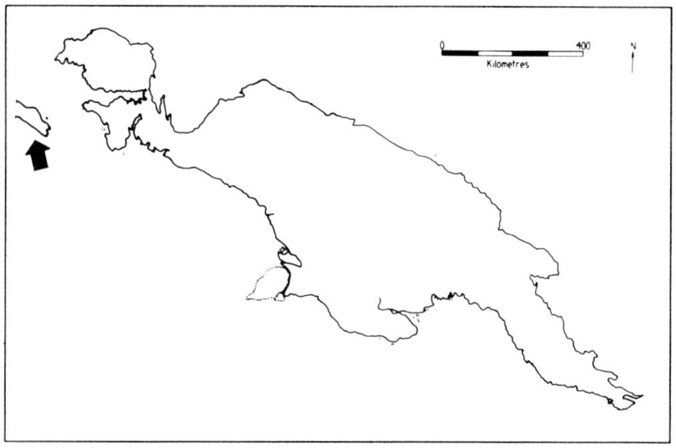

Figure 6. Distribution of the Seram bandicoot, *Rhynchomeles prattorum.*

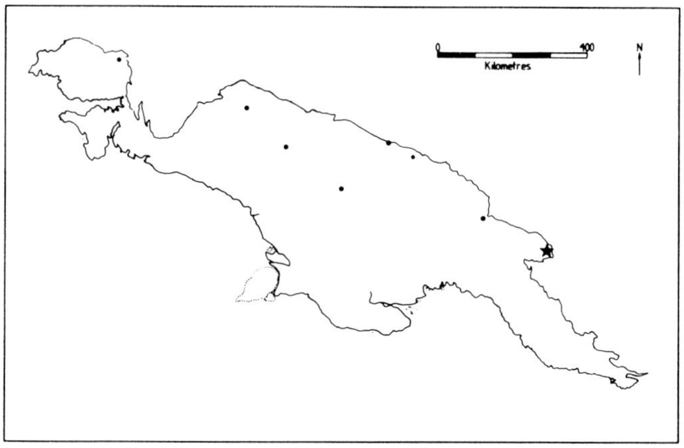

Figure 9. Distribution of the black-spotted cuscus, *Spilocuscus rufoniger.*

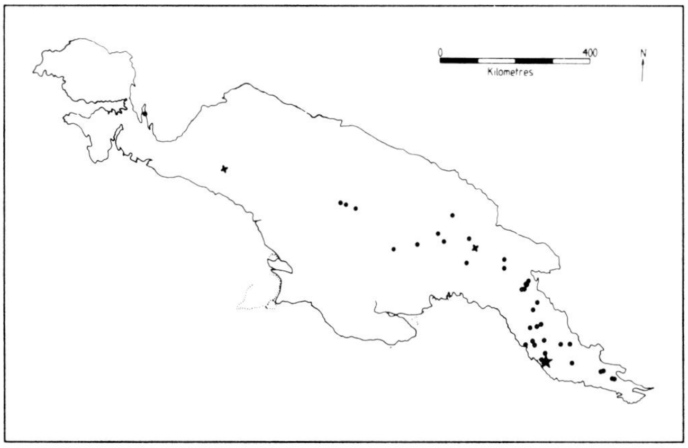

Figure 10. Distribution of the Doria's tree-kangaroo, *Dendrolagus dorianus*.

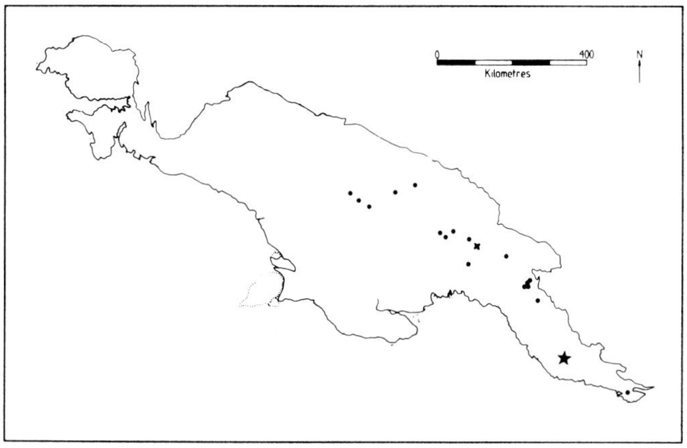

Figure 11. Distribution of the Goodfellow's tree-kangaroo, *Dendrolagus goodfellowi*.

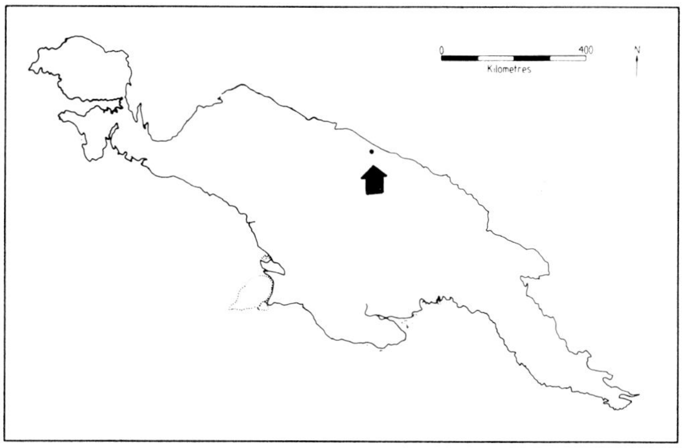

Figure 12. Distribution of the Scott's tree-kangaroo, *Dendrolagus scottae*.

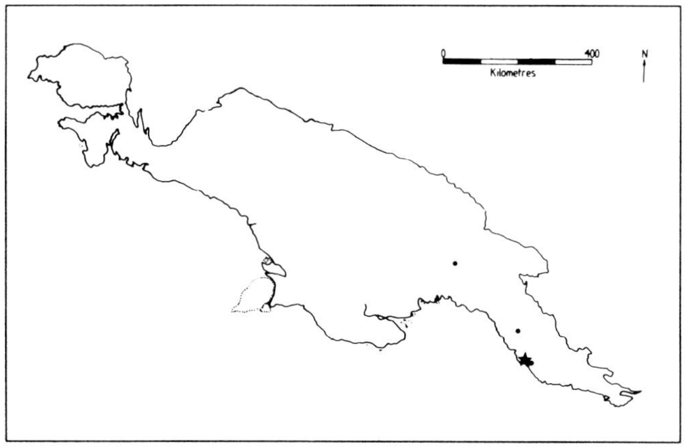

Figure 13. Distribution of the Macleay's dorcopsis, *Dorcopsulus macleayi*.

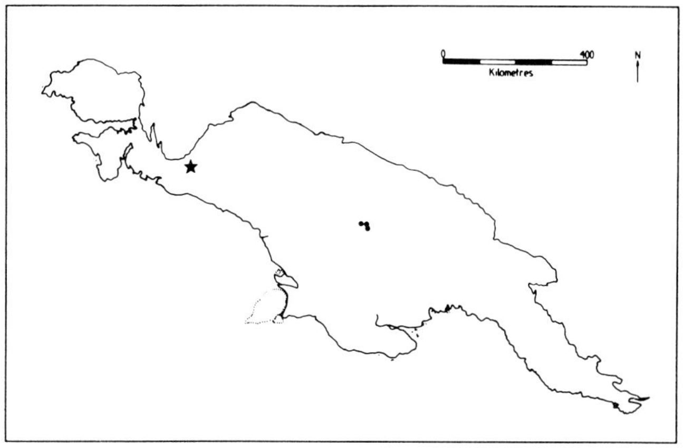

Figure 14. Distribution of the great-tailed triok, *Dactylopsila megalura*.

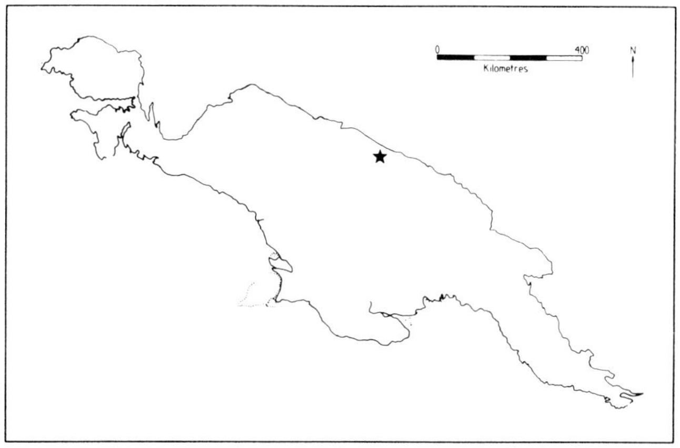

Figure 15. Distribution of the northern glider, *Petaurus abidi*.

4. Conservation of Marsupials and Monotremes in Australia

Australia accounts for about half the world's mammals that have become extinct in modern times. Within the marsupials, ten species (7% of the total) are presumed to be extinct (Table 15) and many other species have declined drastically in range and abundance. Neither species of Australian monotreme is considered to be in immediate danger of extinction.

Australian mammal extinctions may have been even more extreme than current statistics indicate, as some species probably became extinct before being recorded by naturalists or before their full geographic ranges had been delineated. The central hare-wallaby (*Lagorchestes asomatus*), for instance, is known to science from only a single skull, but apparently had a widespread arid zone distribution within living memory (Burbidge et al. 1988). Sub-fossil deposits have revealed several mammal species that probably were extant at European settlement, including an un-named Potoroid from the Nullarbor Plain (Baynes 1987).

Table 15. Extinct species of Australian marsupials (from Burbidge and McKenzie 1989).

Thylacine	*Thylacinus cynocephaus*
Desert bandicoot	*Perameles eremiana*
Pig-footed bandicoot	*Chaeropus ecaudatus*
Lesser bilby	*Macrotis leucura*
Broad-faced potoroo	*Potorous platyops*
Desert rat-kangaroo	*Calyoprymnus campestris*
Eastern hare-wallaby	*Lagorchestes leporides*
Central hare-wallaby	*Lagorchestes asomatus*
Crescent nailtail wallaby	*Onychogalea lunata*
Toolache wallaby	*Macropus greyi*

In recent years there have been varying assessments of the number of marsupial species thought to be in danger of or vulnerable to extinction. However, the two most recent authoritative ones provide similar assessments, viz. Burbidge and McKenzie (1989): 27 species, and CONCOM (Council of Nature Conservation Ministers) (1991): 27 species.

The above national estimates do not show the degree to which increasing numbers of species are under pressure regionally. Marsupial declines have occurred at different rates in different parts of Australia and this is discussed on page 30.

Some of the declined species have suffered enormous changes in range and abundance. The burrowing bettong (*Bettongia lesueur*), for example, once inhabited more than 60% of the Australian mainland, but now occurs only on three islands off the Western Australian coast. Other declines have been similarly dramatic: chuditch (*Dasyurus geoffroii*) from about 60% to less than 1%; numbat (*Myrmecobius fasciatus*) from about 25% to less than 1%; woylie (*Bettongia penicillata*) from more than 60% to less than 1%; and mala (*Lagorchestes hirsutus*), from about 25% to a minute speck of the Tanami Desert and two small islands.

While it is clear that there is major cause for concern about the survival of Australian marsupials, there have been considerable advances in marsupial conservation in recent years. Knowledge of species' ecology is growing rapidly, there is a developing science of endangered species management, and there is heightened interest in conservation in the country. Federal, state, and territory nature conservation agencies are embarking on recovery plans that include habitat protection and management, exotic predator control, and reintroductions.

This Action Plan lists 19 endangered and eight vulnerable marsupial species in Australia that require some form of research or management response; this is 19% of Australian marsupial species (or 26% if we include the ten presumed extinct species). The figure increases yet again to 49% if we take account of a

Boodie/burrowing bettong, *Bettongia lesueur* (photo by Andrew Burbidge, Department of Conservation and Land Management, Western Australia).

further 33 marsupials, listed in this plan as "Potentially Vulnerable" and deemed to be in need of conservation monitoring. Table 16 shows the list of threatened species identified by this Action Plan.

Historical Causes of Decline and Extinction

Before successful management programs can be applied to a threatened species, the factors that caused the initial or continuing decline must be identified, and counter methods devised. For many species, there are insufficient biological and historical data currently available to identify confidently the reasons for their threatened status, and a recent trend has been to examine whole faunas rather than wait for detailed single species studies to be carried out. Burbidge et al. (1988); Burbidge & McKenzie (1989); Johnson et al. (1989); and Morton (1990) all show the possible contributing causal factors, and how singly or in powerful combination, they have led to the extinction or decline of marsupial species.

Mountain pygmy-possum, *Burramys parvus* (photo by J.H. Seebeck, Victorian Department of Conservation and Environment).

Critical Weight Range Species

Many authors have noted that, in Australia, the non-flying, medium-sized mammals have been particularly affected since European settlement (e.g. Calaby 1971, Morton and Baynes 1985; Giles and Lim 1987, Morton 1990). Burbidge and McKenzie (1989) have shown that extinctions and declines are virtually confined to non-flying mammals with mean adult body weights between 35 g and 5,500 g. Variation in patterns of attrition within this critical weight range (CWR) can be explained almost entirely by a combination of regional rainfall and, to a lesser extent, species' habitat and dietary preferences. CWR mammals in greatest danger of extinction are from arid and semi-arid areas. Those species confined to the ground's surface are in most danger and herbivores and omnivores are more likely to become extinct than carnivores. CWR mammals that have declined or become extinct are not restricted to one phylogenetic group but include most groups of marsupials and native rodents.

One hundred and four species of the total Australian mammal fauna of about 260 non-marine species lie within the CWR; these include both monotremes and 80 species of marsupials. Eight of the ten extinct marsupials were CWR species. Of the 27 marsupials listed as "Endangered" or "Vulnerable" in this Action Plan, all but two lie within the CWR (see Chapter 6, table 21).

Causal Factors in Historical Decline

There have been many hypotheses put forward to explain the reasons why particular species or suites of species have declined in, or disappeared from, various parts of Australia. These include clearing for agriculture, draining and salination of wetlands, grazing and browsing by introduced animals, changed fire regimes, introduced predators, disease, and overkill by hunters. It is not easy to measure the extent or importance of these disturbances, especially from the point of view of the native mammals. However, some of these hypotheses can be dismissed as common causes of the disappearance or decline of an array of species throughout their former ranges.

Hypotheses that can be eliminated as general causes of extinction or decline of suites of species:

- **Disease:** There is no direct evidence that disease has led to the extinction of any Australian mammal, although epidemics have been blamed for the decline of carnivorous marsupials in southeastern Australia and Tasmania around the turn of the century (Strahan 1983). It could be argued that disease removed mammals a long time ago when there were no biologists to study the effects, but if disease was a primary cause of extinction or decline in a significant number of species, it is curious that it should have selectively affected the CWR species from a wide variety of phylogenetic groups, especially in arid regions (Burbidge & McKenzie 1989, Johnson et al. 1989).

- **Overkill:** While many species of Australian mammals have been hunted for thousands of years by Aborigines and many were hunted for food by early European settlers, and while hunting of some species continues today, there is no evidence that post-European declines and extinctions have been the direct result of excessive harvesting. Most species exterminated by overkill elsewhere in the world were large (Diamond 1984); in Australia, by contrast, larger species (>5.5 kg), such as kangaroos, have suffered no major declines except where their habitat has been destroyed, while many smaller (i.e. CWR) species that were never hunted by Europeans have disappeared (Burbidge & McKenzie 1989).

Hunting may have contributed to the extinction of the Toolache wallaby (*Macropus greyi*), but habitat destruction (see below) is believed to have been the major cause of its extinction (Robinson and Young 1983). Hunting has also been blamed for the extinction of the thylacine in Tasmania, but the actual cause is open to conjecture.

Table 16. Threatened Australian marsupials and monotremes.

Common Name	Scientific Name	Status
Platypus	*Ornithorhynchus anatinus*	Potentially Vulnerable
Numbat	*Myrmecobius fasciatus*	Endangered
Kultarr	*Antechinomys laniger*	Potentially Vulnerable
Atherton antechinus	*Antechinus godmani*	Potentially Vulnerable
Kowari	*Dasycercus byrnei*	Endangered
Mulgara	*Dasycercus cristicauda*	Vulnerable
Chuditch/Western quoll	*Dasyurus geoffroii*	Endangered
Tiger quoll	*Dasyurus maculatus*	Potentially Vulnerable
Eastern quoll	*Dasyurus viverrinus*	Potentially Vulnerable
Dibbler	*Parantechinus apicalis*	Endangered
Carpentarian antechinus	*Parantechinus mimulus*	Potentially Vulnerable
Little red antechinus	*Parantechinus rosamondae*	Potentially Vulnerable
Red-tailed phascogale	*Phascogale calura*	Endangered
Brush-tailed phascogale	*Phascogale tapoatafa*	Potentially Vulnerable
Kangaroo Island dunnart	*Sminthopsis aitkeni*	Potentially Vulnerable
Carpentarian dunnart	*Sminthopsis butleri*	Potentially Vulnerable
Julia Creek dunnart	*Sminthopsis douglasi*	Endangered
Sandhill dunnart	*Sminthopsis psammophila*	Vulnerable
Red-cheeked dunnart	*Sminthopsis virginiae*	Potentially Vulnerable
Golden bandicoot	*Isoodon auratus*	Endangered
Southern brown bandicoot	*Isoodon obesulus*	Potentially Vulnerable
Western barred bandicoot	*Perameles bougainville*	Endangered
Eastern barred bandicoot	*Perameles gunnii*	Vulnerable
Bilby/Dalgyte	*Macrotis lagotis*	Vulnerable
Rufous spiny bandicoot	*Echymipera rufescens*	Potentially Vulnerable
Marsupial mole	*Notorcytes typhlops*	Potentially Vulnerable
Koala	*Phascolarctos cinereus*	Potentially Vulnerable
Northern hairy-nosed wombat	*Lasiorhinus krefftii*	Endangered
Northern brushtail possum	*Trichosurus arnhemensis*	Potentially Vulnerable
Mountain brushtail possum	*Trichosurus caninus*	Potentially Vulnerable
Common brushtail possum	*Trichosurus vulpecula*	Potentially Vulnerable
Rufous bettong	*Aepyprymnus rufescens*	Potentially Vulnerable
Tasmanian bettong	*Bettongia gaimardi*	Potentially Vulnerable
Boodie/Burrowing bettong	*Bettongia lesueur*	Endangered
Woylie/Brush-tailed bettong	*Bettongia penicillata*	Endangered
Northern bettong	*Bettongia tropica*	Endangered
Musky rat-kangaroo	*Hypsiprymnodon moschatus*	Potentially Vulnerable
Long-footed potoroo	*Potorous longipes*	Endangered
Long-nosed potoroo	*Potorous tridactylus*	Potentially Vulnerable
Spectacled hare-wallaby	*Lagorchestes conspicillatus*	Potentially Vulnerable
Mala/Rufous hare-wallaby	*Lagorchestes hirsutus*	Endangered
Banded hare-wallaby	*Lagostrophus fasciatus*	Endangered
Tammar wallaby	*Macropus eugenii*	Vulnerable
Western brush wallaby	*Macropus irma*	Potentially Vulnerable
Bridled nailtail wallaby	*Onychogalea fraenata*	Endangered
Narbarlek	*Petrogale concinna*	Potentially Vulnerable
Godman's rock-wallaby	*Petrogale godmani*	Potentially Vulnerable
Black-footed rock-wallaby	*Petrogale lateralis*	Vulnerable
Brush-tailed rock-wallaby	*Petrogale penicillata*	Vulnerable
Proserpine rock-wallaby	*Petrogale persephone*	Endangered
Yellow-footed rock-wallaby	*Petrogale xanthopus*	Potentially Vulnerable
Quokka	*Setonix brachyurus*	Potentially Vulnerable
Mountain pygmy-possum	*Burramys parvus*	Endangered
Little pygmy-possum	*Cercartetus lepidus*	Potentially Vulnerable
Greater glider	*Petauroides volans*	Potentially Vulnerable
Western ringtail	*Pseudocheirus occidentalis*	Vulnerable
Leadbeater's possum	*Gymnobelideus leadbeateri*	Endangered
Yellow-bellied glider	*Petaurus australis*	Potentially Vulnerable
Squirrel glider	*Petaurus norfolcensis*	Potentially Vulnerable
Honey-possum	*Tarsipes rostratus*	Potentially Vulnerable

- **Clearing for agriculture and other human activities:** Clearing has been confined to relatively small parts of Australia and few if any mammals were restricted to areas that were cleared at the time they became extinct; most species extended from areas that are now the agricultural districts into adjacent semi-arid and arid areas, and they have disappeared from these as well. While clearing has doubtless been responsible for the local decline and disappearance of several mammal species, it could not have been the primary cause of extensive decline for many.

  Likewise, drainage and salination could not have affected many species since few were restricted to areas that have been so affected.

  The only exception was one of the two non-CWR species that is extinct—the Toolache wallaby, which occurred in the southeast of South Australia and the southwest of Victoria. Robinson and Young (1983) have reported that swamps formed a significant part of its habitat and that most of them have now been drained in a series of schemes that commenced as early as 1862. They concluded that destruction of habitat, including clearing of vegetation and drainage, was the major cause of the demise of the species (Burbidge & McKenzie 1989).

The remaining hypotheses:

The remaining possible causes—introduced herbivores, changed fire regimes, and exotic predators—are extensive in their distribution and effect. Burbidge and McKenzie (1989) concluded that environmental changes since European settlement have emulated an increase in aridity by reducing the environmental productivity available to vertebrates. These changes include the diversion of environmental resources to humans and introduced mammals, and a reduction in vegetative cover by exotic herbivores and changed fire regimes. Their analyses supported the view that the reduction in available productivity has caused CWR mammals to suffer the greatest attrition because of their limited mobility but relatively high daily metabolic requirements. The direct elimination of confined populations of mammals by exotic predators has exacerbated this attrition.

- **Exotic Herbivores:** Exotic herbivores—rabbits, goats, cattle, sheep, donkeys, pigs, camels, and so on—are widespread in Australia. Pastoralism is the major land-use of much of the arid zone, but many feral species extend beyond leased areas into conservation reserves and unoccupied deserts. The effects of introduced herbivores on the native vegetation have been significant and their effects on the habitats of the indigenous mammals that depend on the vegetation for food and shelter have also been significant. Burbidge & McKenzie (1989) reported that there are no data to suggest that introduced herbivores have had an appreciable effect on the spinifex communities that dominate the sandy and stony deserts, suggesting that introduced herbivores were not the sole cause of mammal extinctions and declines. Morton (1990), however, has suggested that many arid zone mammals depended, in times of drought, on small areas of habitat

Chuditch/Western quoll, *Dasyurus geoggroii* (photo by Keith Morris-Department of Conservation and Land Management, Western Australia).

  of exceptionally high quality (in terms of the dependability and nutritional character of their plant growth), and that these areas have been degraded by exotic herbivores, especially rabbits.

- **Fire regimes:** Fire is one of the most significant modifiers of habitat in Australia, but while there are several studies on the effects of single fires on mammal populations, there are few data on the effects of changed fire regimes on the status of CWR mammals.

  Fire regimes are known to have changed significantly in parts of Australia since European settlement. The propensity for aborigines to burn the country was the subject of comment by some of the earliest Europeans to visit Australia (see Nicholson 1981 for a review) and many explorers and settlers recorded the interaction of fire, plants, and mammals (Johnson et al. 1989). In much of the western deserts the movement of aborigines to settlements has occurred only in the last 25 to 60 years, and it is possible to reconstruct crudely the pre-European situation. Here the fire regime changed from one where mainly small fires occurred throughout the year to one dominated by infrequent, large, summer fires that have a homogenizing effect on the vegetation, which has changed from one of small patches in different stages of recovery following fire to one with much larger, less diverse patches (Latz and Griffin 1978, Kimber 1983). Burbidge et al. (1988) have shown that the disappearance of mammals from the deserts of Western Australia, northwestern South Australia, and southwestern Northern Territory coincided with the depopulation of the area by aborigines and the subsequent change in fire regime.

- **Introduced predators:** Three exotic predators are now widespread in Australia. The dingo (*Canis familiaris dingo*) is thought to have been introduced between 3,000 and 8,000 years ago (Newsome 1983), and it is usually assumed that any changes its colonization had on the status of mammals predated European settlement. In contrast, the feral cat (*Felis* *catus*) and the red fox (*Vulpes vulpes*) are much more recent arrivals.

The cat occurs throughout the continent and appears to have been present for a long time. Burbidge et al. (1988) report desert aborigines' statements that cats have "always been present" and that they "arrived from the west". European explorers in the western deserts in the 1890s found cats to be present and a common aboriginal food item. Cats could have become established in Australia following European settlement on the east coast, but it is possible that they arrived earlier from shipwrecks on the west coast in the 17th century or from visits by Asian fishermen well before European settlement. Cats appear to have eliminated populations of CWR mammals from some offshore islands (Burbidge and McKenzie 1989) and there is no doubt that cats prey upon CWR marsupials as large as mala (or rufous hare-wallabies) (Johnson et al. 1989). However, extinct CWR marsupials such as desert rat-kangaroo (*Caloprymnus campestris*, 890 g), central hare-wallaby (*Lagorchestes asomatus*, ca. 1,500 g) and pig-footed bandicoot (*Chaeropus ecaudatus*, ca. 200 g) persisted until at least the 1930s and probably as late as the 1950s (Burbidge et al. 1988), suggesting that the cat has not been the primary factor in their declines.

The fox arrived much later. Although the earliest release may have been in 1845, Coman (1983) places the time of establishment in the 1860s and the place as Victoria. It spread rapidly, colonizing most of the continent except the wet tropics by the 1930s. The lack of mammal extinctions in the wet tropics is consistent with the distribution of the fox, and foxes have been implicated in the demise of many native Australian mammals, but the relationship is not absolute; in many parts of Australia it is clear that the native mammals had begun to decline or had disappeared before foxes arrived (Burbidge et al. 1988, Johnson et al. 1989).

Frankel and Soulé (1981) state that the evidence for an exotic predator (other than *Homo sapiens*) causing the extinction of native species on continents anywhere in the world is unconvincing. However, this statement has proved to be untrue in Australia, where in the last 20,000 to 30,000 years marsupials have not been challenged by a fox-sized predator (Johnson et al. 1989).

Recent work has clearly demonstrated that foxes can eliminate remnant CWR mammal populations on both the mainland (Kinnear et al. 1988) and islands (Burbidge and McKenzie 1989) and control of foxes is a pre-requisite for both the management of remnant CWR mammal populations and the successful re-introduction of CWR species to parts of their former range. There is now much speculation that foxes have been the primary cause of the extinction and decline of many CWR species.

Thus, it appears that the interaction of three factors—changes to habitat caused by introduced herbivores, homogenization of habitat following changed fire regimes, and the spread of exotic predators—has been responsible for the high extinction rate since European settlement of Australia.

Regional Differences in Marsupial Decline

Mainland Australia

Marsupial declines have occurred at different rates in different parts of Australia. The greatest decline has occurred in the arid regions; approximately 33% of mammal species are locally extinct in sandy and stony desert ecosystems (Burbidge et al. 1988, Burbidge & McKenzie 1989), including six of the ten extinct marsupials. This is the highest regional extinction rate in the country and it has occurred without impact by agriculture, pastoralism, or mining on most of this very large region.

Because of their relatively simple ecosystems and the few disturbances to them, the deserts are particularly amenable to studies aimed at elucidating causes of extinction (Burbidge & McKenzie 1989, Johnson et al. 1989, Morton 1990). These studies have been aided by the experience and expertise of the aborigines of the area, many of whom knew and hunted the now-extinct species and who are able to recount both the ecology of the mammals and the history of environmental changes and extinctions (Burbidge et al. 1988). In the arid regions, the major factor causing the extinction of most of the CWR species has been habitat change, brought about by introduced herbivores and

Woylie/Brush-tailed bettong, *Bettongia penicillata* (photo courtesy of WWF Australia).

changes to fire regimes, exacerbated by the effect of introduced predators on remnant populations.

In semi-arid lands there have also been high levels of extinction and decline. Heavily cleared semi-arid lands have been hard hit: in the Western Australian wheatbelt, for example, from a total mammal fauna of 56 species (32 marsupials), 14, including nine marsupials, are locally extinct and 12, including 11 marsupials, are threatened with extinction (Kitchener et al. 1980, Burbidge & McKenzie 1989).

The marsupial faunas of the wet tropics are largely intact, while those of the moist temperate zones have suffered intermediate declines. In the wetter temperate areas of the southeast, the effects of introduced herbivores and fire have not been as marked. The loss of the eastern quoll, small macropodoids, and the decline of bandicoots are most likely due to the combination of habitat change brought about by agriculture and exotic predators, possibly exacerbated by disease. In the southwest, CWR species have also declined considerably because of habitat change, but their decline has been buffered by the presence of vegetation containing natural fluoro-acetates, a powerful toxin, to which the local indigenous populations have an evolved immunity not shared by exotic predators. Rainforests, which are of extremely limited extent in Australia, are of great importance to the conservation of a number of relict marsupial species.

Table 17. Australian mainland marsupials that now occur only on islands.

Species	Island(s)
Dasyurus viverrinus (Eastern quoll)	Tasmania
Pseudantechinus mimulus (Carpentarian antechinus)	Centre, North & South-west, Sir Edward Pellew Group, N.T.
Perameles bougainville (Western barred bandicoot)	Bernier & Dorre, W.A.
Bettongia lesueur (Burrowing bettong)	Barrow, Bernier & Dorre, W.A.
Bettongia gaimardi (Tasmania bettong)	Tasmania, Bruny
Lagostrophus fasciatus (Banded hare-wallaby)	Bernier & Dorre, W.A.
Thylogale billardierii (Tasmanian pademelon)	Tasmania & 16 other Tasmanian Islands

The Importance of Islands

At least 62 species of marsupials (46 of which are CWR species) occur or occurred on 160 islands around the Australian continent (see Table 18 for complete listing for Endangered, Vulnerable, and Potentially Vulnerable species on Australian islands). Seven species of mainland marsupials now occur only on islands (Table 17). These figures alone convey the immensely valuable role that islands are playing and must continue to play in protecting marsupial species.

Some of these islands have been affected by a number of the causal factors associated with declines and extinctions on the mainland, for instance: introduced herbivores and carnivores such as goats, sheep, pigs, foxes, cats and rats; tourism and other recreation; mining; timber extraction; military uses; and fire (Burbidge 1989). Most, however, are unaffected and remain essential refugia for some critically threatened species.

Tasmania became a refuge for the Tasmanian devil (*Sarcophilus harrisii*) and the thylacine (*Thylacine cynocephalus*) when the dingo (*Canis familiaris*) replaced them on the mainland. With European settlement, Tasmania became a refuge for the Tasmanian (red-bellied) pademelon (*Thylogale billardierii*), Tasmanian (eastern) bettong (*Bettongia gaimardi*), and eastern quoll (*Dasyurus viverrinus*), all of whose mainland populations are now extinct. Although habitat change is putting pressure on several marsupial species, the fox is not established there. The effects of feral cats in Tasmania have not been studied.

As Tables 17 and 18 show, the Western Australian islands of Barrow, Bernier, and Dorre are of the highest conservation importance, harbouring four of Australia's most threatened marsupial species, the western barred bandicoot (*Perameles bougainville*), the burrowing bettong (*Bettongia lesueur*), the mala or rufous hare-wallaby (*Lagorchestes hirsutus*), and the banded hare-wallaby (*Lagostrophus fasciatus*). These and other surviving island populations also provide a great opportunity to conduct further research on island biogeography, ecology, and threatened species biology, in preparation for future reintroduction programs on the mainland. A high level of research and management commitment and effort will be required to take advantage of this opportunity.

Distribution of Endangered and Vulnerable Australian Marsupials

Records since 1980 of the 27 Endangered and Vulnerable Australian marsupials were plotted onto a map of Australia with a 1° latitude and 1.5° longitude grid, i.e. the boundaries of the 1:250 000 map series. Table 19 shows the number of map sheets per species and Figure 16 shows the distribution of these threatened taxa. It can be seen from Table 19 that one species has not been recorded at all since 1980, five species are restricted to a single map sheet and another four are restricted to two sheets. At the other extreme, the bilby occurs in 36 map sheets in two states and the Northern Territory, while the brush-tailed rock-wallaby occurs in 37 map sheets covering three states. (The Australian Capital Territory was not considered separately from New South Wales for the purposes of this exercise.)

The number of map sheet records per state was Western Australia: 92, Northern Territory: 37, Queensland: 43, South Australia: 16, New South Wales: 16, Victoria: 4, Tasmania: 4.

These statistics show that the greatest concentration of endangered and vulnerable marsupials is in the south west of Western Australia and Shark Bay, and that Western Australia (16 species) and Queensland (10 species) have the greatest number of threatened marsupial species within their borders.

Bilby/Dalgyte, *Macrotis lagotis* (photo by R. Southgate, Conservation Commission of the Northern Territory).

quirements of the Victorian government's Flora and Fauna Guarantee Act, while an "Action Plan" has also been derived from the Management Plan for the mountain pygmy-possum.

There are a number of other marsupial recovery plans that are currently near to or in full operational mode: mala, *Lagorchestes hirsutus* (E) and bilby, *Macrotis lagotis* (V) (Conservation Commission of the Northern Territory); bridled nailtail wallaby, *Onychogalea fraenata* (E), and northern hairy-nosed wombat, *Lasiorhinus krefftii* (E) (Queensland National Parks and Wildlife Service); numbat, *Myrmecobius fasciatus* (E) (Department of Conservation and Land Management, Western Australia); brush-tailed bettong, *Bettongia penicillata* (E) (South Australian National Parks and Wildlife Service); and forester kangaroo, *Macropus giganteus tasmaniensis* (PV) (Department of Parks, Wildlife and Heritage, Tasmania).

Draft management plans have been prepared for Leadbeater's possum, *Gymnobelideus leadbeateri* (E) and koala, *Phascolarctos cinereus* (AS) in Victoria, and plans are currently being prepared for the koala (PV) and mountain pygmy-possum (E) in New South Wales; for the koala (PV) and bilby (E) in Queensland; for the black-footed rock-wallaby, *Petrogale lateralis* (V) in Western Australia; for the common brushtail possum (E), Mulgara *Dasycercus cristicauda* (V), and nabarlek, *Petrogale concinna* (PV) in the Northern Territory, and for the Tasmanian bettong, *Bettongia gaimardi* (PV) in Tasmania.

New South Wales National Parks and Wildlife Service has also prepared, but not yet published or implemented, management plans for the platypus *Ornithorhynchus anatinus* (PV), the yellow-footed rock-wallaby, and arid zone dasyurids.

Two national organisations, one government and one non-government, have also contributed substantially to marsupial and monotreme research and management. The Australian National Parks and Wildlife Service (ANPWS) and World Wide Fund for Nature Australia have allocated funds and/or expertise to a large number of the previously mentioned species management activities, and continue to make such commitments.

It has been noted that most wildlife conservation agencies are now embarking on recovery plans for threatened marsupial species. The administrative and legal arrangements required to permit such developments over the longer-term are also being established by many states and territories.

In Victoria, the government has introduced a "Flora and Fauna Guarantee Act", which aims to protect the breadth of that state's biodiversity, including the development of recovery programs for

Table 19. Number of map sheets per threatened species.

No.	Species	State(s)
0	Julia Creek dunnart	QLD
1	Leadbeater's possum	VIC
	Mountain pygmy-possum	VIC & NSW
	Northern hairy-nosed wombat	QLD
	Proserpine rock-wallaby	QLD
	Bridled nailtail wallaby	QLD
2	Golden bandicoot	WA
	Western barred bandicoot	WA
	Long-footed potoroo	VIC & NSW
	Banded hare-wallaby	WA
3	Dibbler	WA
	Boodie	WA
	Northern bettong	QLD
	Mala	WA & NT
4	Sandhill dunnart	WA & SA
5	Eastern barred bandicoot	VIC & TAS
	Western ringtail	WA
6	Numbat	WA
	Red-tailed phascogale	WA
	Kowari	QLD & SA
9	Brush-tailed bettong	WA & SA
11	Tammar wallaby	WA & SA
14	Chuditch	WA
19	Mulgara	WA, NT, SA & QLD
34	Black-footed rock-wallaby	NT, WA & QLD
36	Bilby	WA, NT & QLD
37	Brush-tailed rock-wallaby	NSW, VIC & QLD

Table 20. Marsupials for which some stage of recovery plans are being written for the Endangered Species Program, 1990/91 (lead agency is indicated).

Queensland National Parks and Wildlife Service	
Sminthopisis douglasi	Julia Creek dunnart
Macrotis lagotis	Bilby
Lasiorhinus krefftii	Northern hairy-nosed wombat
Onychogalea fraenata	Bridled nailtail wallaby
Bettongia tropica	Northern bettong
Petrogale persesphone	Proserpine rock-wallaby
South Australian National Parks and Wildlife Service	
Dasyuroides byrnei	Kowari
New South Wales National Parks and Wildlife Service	
Burramys parvus	Mountain pygmy-possum
Victorian Department of Conservation & Environment	
Potorous longipes	Long-footed potoroo
Petrogale penicillata	Brush-tailed rock-wallaby
Tasmanian Department of Parks, Wildlife and Heritage	
Perameles gunnii	Eastern barred bandicoot
Western Australia Department of Conservation and Land Management	
Dasyurus geoffroii	Chuditch
Bettongia penicillata	Brush-tailed bettong
Petrogale lateralis	Black-footed rock-wallaby (WA races only)
Macropus eugenii	Tammar wallaby
Conservation Commission of the Northern Territory	
Macrotis lagotis	Bilby
Lagorchestes hirsutus	Mala
Dasycercus cristicauda	Mulgara

threatened species. Legislation for the protection of threatened species and habitats is also in the process of development in New South Wales, Tasmania, Western Australia, and Queensland, while the Commonwealth government is likely to introduce such legislation in 1992.

At the Federal level, the government has established an Endangered Species Program (ESP), that has now developed a "National Strategy for the Conservation of Species and Habitats Threatened with Extinction" (ANPWS 1989). This program allocates most of its resources to the research and management of threatened species, including many marsupials. Table 20 shows the Action Plan species currently benefiting from ESP resources, complementing the work of State and Territory agencies. The ESP is likely to fund further marsupial recovery plans during the 1991-1992 financial year.

Zoos have also demonstrated a capacity to support a range of endangered species programs through maintenance and breeding of animals in captivity and as a resource for a range of research activities. Australian zoological institutions are currently collaborating with wildlife agencies in several programs on endangered Australian fauna. Many Australasian marsupials and monotremes, including most genera, have been bred in captivity (George 1990). The platypus cannot readily be captive-bred, and wombats cannot be bred reliably. These species will probably yield to behavioural and physiological research, of which little has been done to date. The current capacity of Australian zoos to contribute to captive breeding programs is limited by the number of spaces available for new stock, whether obtained from the wild or bred in captivity, and by staff and finance. Such limitations could be severe if a decision is taken to maintain one or more species entirely in captivity due to a hopeless situation with the wild population. This problem is, however, less severe with the majority of Australasian fauna than it is with the larger species on other continents (J. Giles, pers. comm.).

A relatively large number of spaces are available in overseas zoos and there is considerable interest in obtaining and breeding Australasian species. These spaces may become very valuable in cases where a decision is taken to rely entirely upon captive breeding, even for a relatively small number of years, to ensure the survival of a species. However, reintroduction of marsupials and monotremes bred overseas, particularly for release programs, is problematical because of poor information on the susceptibility of these species to sylvatic diseases currently exotic to Australia (J. Giles, pers. comm.).

Table 21, Chapter 6 provides the captive breeding classifications under the "Australasian Species Management Program" of the major zoos, and indicates next to each species the status of such management programs if any. It should be noted that classifications represent the status within zoos' captive breeding programs, not necessarily representing national or state conservation priorities. Several critical species are due for reclassification as regional captive breeding programs are developed in accordance with conservation priorities (G. George, pers. comm.).

Kowari, *Dasycercus byrnei* (photo courtesy of the South Australian Museum).

5. Conservation Management of Small Populations

Graeme G. George[1] and Peter R. Brown[1]

Adaptation and Evolutionary Potential

The ultimate goal for single species conservation programs is the restoration of viable populations in natural habitat, though this may be well into the future for many species.

Under natural conditions, most species have the capacity to evolve in response to environmental change. Populations that are unable to respond to changed conditions go extinct. Adaptive change occurs when environmental selection pressures act upon genetic differences among individuals within a population. This accumulated genetic variability needs to be preserved if a threatened species is to be conserved with some long-term evolutionary potential (Franklin 1980, Soulé 1980, Frankel & Soulé 1981, Soulé et al. 1986, Allendorf & Leary 1986).

Genetic variation includes two components: *allelic variation* across the population, and *heterozygosity* of individuals. The presence of multiple variants (alleles) of genes imparts the inter-individual variation on which natural selection acts to produce adaptive evolution. The presence of non-identical alleles at genetic loci within an individual (heterozygosity) is often associated with individual fitness, especially in stressful environments. Therefore, while immediate fitness of individuals often depends on their heterozygosity, the long-term evolutionary flexibility of the population is determined by allelic variation.

Small populations lose allelic variation through the randomness of Mendelian inheritance (genetic drift). In a large breeding population, allele loss is normally balanced, over time, by the flow of alleles from individuals who disperse to breed outside their natal areas. With input from new mutations, there is no net loss of variation in the gene pool. In small populations, the negative effect of random changes to gene frequencies is much greater than the positive effects of selection (Frankel & Soulé 1981). Thus populations that have lost allelic diversity due to drift are unlikely to have the resilience to cope with environmental challenges.

Eastern barred bandicoot, *Perameles gunnii* (photo by J.H. Seebeck, Victorian Department of Conservation and Energy).

Though homozygosity may not be a problem in a stable environment, few threatened species are in such a situation. Numerically, small populations tend to lose both allelic variation and heterozygosity, and recovery plans need to take this into account as these losses can be expected to have profound effects on the probability of recovery and survival.

The Extinction Process

In the conservation context, extinction is best viewed as a population-level process, not an event (Frankel & Soulé 1981, Gilpin & Soulé 1986). Clark et al. (1990) describe a general extinction model illustrating the role of systematic and random factors operating on fragmented populations. During the early stage of decline, systematic (deterministic) factors, such as habitat loss, can be overcome by preventative management. However, if deterministic factors continue to reduce the size of isolated populations, random factors, such as drought or wildfire, begin to influence the probability of their survival. Such isolated sub-populations, within a fragmented meta-population, will have varying probabilities of survival (Gilpin 1987), but with the loss of each isolate, the gene pool of the species is diminished, and each loss becomes a step towards the species extinction. In terms of the Clark model we have, in most cases, lost the opportunity for preventative management of sub-populations of many of our threatened marsupials, and are now at the stage of crisis management of very small remnant isolates.

Low population size is a characteristic of most threatened species. Small populations may occur at low densities over a wide area of suitable habitat, as with many rare predators, or may consist of localised populations confined to remnant pockets. Isolation may result from the fragmentation of once-continuous habitat by human activities, or from the elimination of intermediate populations by overexploitation. Where populations become

[1]Deakin University, Rusden Campus, Clayton, Victoria, Australia.

Mala/Rufous hare-wallaby, *Lagorchestes hirsutus* (photo by Ken Johnson, Conservation Commission of the Northern Territory).

isolated, by loss of habitat without corridors to allow migration or dispersal, or by unsustainable harvest rates, normal processes that produce gene flow are interrupted. In addition to the deterministic processes that fragmented their populations in the first place, isolates of non-volant small mammals are particularly vulnerable to a series of threats of a stochastic (random) nature:

- loss of genetic variability due to random drift and loss of fitness due to inbreeding;
- stochastic demographic events, such as skewed sex-ratios (a problem in very small populations);
- stochastic environmental variation, such as weather, disease and predation; and
- catastrophes, such as wildfire or cyclones.

The factors that increase the probability of isolated populations dying out may also prevent or hinder their replacement by natural colonization (Shaffer 1981, Gilpin 1987). Increased levels of inbreeding in small populations, leading to reduction in fitness, further limit the prospects for recovery from catastrophes. Even if a small number of breeders survive an environmental catastrophe, the resulting genetic bottleneck is likely to further limit diversity in the remnant population. For each species there is a threshold for population size below which feedback mechanisms tend to exacerbate the problems caused by genetic, environmental and demographic stochasticity (Gilpin & Soulé 1986). Once a population falls below its critical size (the minimum viable population) extinction becomes inevitable without intervention. The slide towards extinction has been aptly termed the "extinction vortex".

To sustain long-term evolutionary potential, conservation programs need to manage fragmented populations as meta-populations. The expedient option of basing species recovery on the largest or apparently most secure of the sub-populations poses a great risk of irretrievable loss of valuable genetic diversity as the smaller/more isolated/more peripheral populations spiral to extinction.

The eastern quoll (*Dasyurus viverrinus*) illustrates the vulnerability of small isolated populations. In southeastern Australia, a major reduction in populations of this, and several other species, occurred around the turn of the century. However the quoll survived in isolated colonies around Lake Corangamite (western Victoria), Studley Park (Melbourne), and Vaucluse (Sydney) until the 1950s. The species may have survived longer in the New England National Park and at Wilsons Promontory, but there have been no confirmed records from mainland Australia for many years. Though the final cause of demise of each of the quoll isolates is unknown, introduced predators are known to have been responsible for the more recent loss of colonies of the mala (*Lagorchestes hirsutus*) in the Tanami Desert (K. Johnson, pers. comm.), and of the Parma wallaby (*Macropus parma*) to the Illawarra district of New South Wales (J. Giles, pers. comm.).

The Need for Genetic and Demographic Management of Small Populations

Population genetics theory demonstrates the need to manage threatened populations genetically and demographically in addition to the more traditional environmentally-oriented conservation measures—habitat protection and restoration, control of predators, legal protection, etc.

The more restricted a species becomes, in terms of both range and population size, the more actively it needs to be managed, to replace the natural processes that no longer operate due to the direct or indirect impact of human activity. Most reserves are far too small to support mammalian populations large enough to be viable, under stable natural conditions, for more than the short term, and fewer still are large enough to buffer the effects of major environmental disturbance (Shaffer 1981, Frankel & Soulé 1981, Shaffer 1987). As opportunities to expand reserves are very limited in this overcrowded world, genetic and demographic management of threatened species will be essential if gene pools are to be preserved and populations restored to viable levels.

Translocations

The practice of translocating animals from existing populations, assessed as being able to sustain a harvest, and their release directly into suitable habitat at other sites, is used to establish new populations and to reinforce existing populations which are demographically unstable or genetically impoverished.

The establishment of new populations can assist conservation of a threatened taxon by:

- increasing its total population size;
- providing a greater geographic spread in case of catastrophe;
- improving the probability of preserving uncommon alleles; and
- exposing the gene pool to a greater range of selection pressures.

Successful reintroductions to sites within the historical range of a taxon depend on the absence of, or the reduction and management of, the causal factors that eliminated the original population. In Australia this usually involves habitat restoration and removal or control of exotic predators. In Papua New Guinea and the Wallacean Islands of Indonesia, habitat destruction and hunting

Numbat, *Myrmecobius fasciatus* (photo by Dick Whitford/WWF Australia).

pressure are more likely to be the factors needing to be addressed.

Reintroductions may also be used to restore small populations lost to stochastic catastrophes, where natural re-colonization is prevented by habitat fragmentation.

Introduction to sites outside a species' historical range, e.g. to islands, is an alternative that may have short-term appeal to managers where the removal of predators, restoration of habitat, or traditional rights of land use prevent reintroductions. South Australia has pursued this course with the introduction of the brush-tailed bettong (*Bettongia penicillata*) and the stick-nest rat (*Leporillus conditor*) to islands (P. Coney, pers. comm.). The ecological impact of introducing a new species to a community needs to be assessed before this strategy is used, and monitored after reintroduction, as it is inevitable that some community members will be affected as the new species establishes a niche.

Translocation has been a traditional practice among island peoples in tropical Australasia. Many Wallacean and New Guinea island populations of cuscus (*Spilocuscus* spp. and *Phalanger orientalis*) are believed to have been established by seafaring people as a food source, and oral traditions describe the re-population of some islands with cuscus following volcanic eruption in Papua New Guinea. The population of Matschie's tree-kangaroo (*Dendrolagus matschiei*) on Umboi Island is believed to be introduced.

Current conservation programs in Western Australia are using translocation effectively to reintroduce the brush-tailed bettong and numbat (*Myrmecobius fasciatus*) to areas from which they have disappeared. These projects have undertaken predator control (of foxes) to assist establishment and have radio-tracked released individuals to monitor their fate (A.J. Friend, pers. comm).

In establishing new populations, the gene pool of the source population needs to be sampled adequately to ensure that the founding population has sufficient allelic diversity to aid establishment and persistence. It may be desirable to obtain founding stock from more than one source population where fragmentation has increased the probability of genetic erosion through inbreeding and genetic drift, or where the retained diversity in small isolates needs to be preserved by spreading their alleles around. Mixing of gene pools should only be done after assessing whether diversity between populations is due to adaptive selection or random drift. Molecular genetics is helping to determine the evolutionary significance of observed diversity between populations, e.g., the status of the Victorian and Tasmanian populations of the eastern barred bandicoot (*Perameles gunnii*) (George et al. 1990, Robinson et al. 1990).

Reinforcement of introduced and reintroduced populations by periodic translocation of additional animals may be necessary to increase the effective founder number, where insufficient animals are available at the initial establishment or numbers of translocated animals fail to establish. A minimum of 20 founders is recommended to capture around 98% of the heterozygosity in the source population (Soulé et al. 1986, Foose et al. 1986, Lacy 1989) when establishing captive populations. The same criteria should be applied to new wild populations if they are to remain viable, once established, without continual reinforcement. This minimum figure assumes that all founders will contribute their genes adequately to the subsequent population. This is rarely the case. Many released animals do not become founders, that is, they do not breed successfully, and many do not leave an adequate number of offspring. Estimates of the number of offspring required to pass on all alleles present in founding pairs vary from seven to pass on 99% of alleles (Ballou & Foose, in press) to 12 to pass on all alleles at a probability of 99% (Thompson, in press). Where founder number is limited, monitoring of the reproductive success of founders is highly desirable. Supplementation with additional founders, if available, may be indicated. Where fecundity is naturally low, a higher number of animals may be required for initial release to increase the genetically-effective founder number.

Translocation has a further role where natural gene flow is not possible. Managed migration of animals between sub-populations within a meta-population is expected to be necessary to maintain levels of heterozygosity and fitness well into the future. Lacy (1987) demonstrates the genetic advantages of managed migration.

The source population's dynamics will determine whether adult or subadult animals are selected for translocation. Adult animals can be expected to acclimatise and commence breeding at an early date, but their removal may affect reproduction in the source population. The removal of subadult animals, which may face high mortality rates during dispersal from or within the source population, would leave the breeding population intact.

Captive Breeding

The importance of captive breeding as a conservation tool has been recognised by the IUCN, with the adoption in 1987 of a Policy Statement on Captive Breeding, prepared by the Captive

Breeding Specialist Group of the SSC (IUCN 1988). The statement stresses the need to take threatened species into captivity when wild populations are still in their thousands in order to sample adequately the genetic variability characteristic of the taxon. Unfortunately, most recovery programs have treated captive breeding as an option of last resort. Delaying the establishment of a captive breeding population can sometimes result in irretrievable loss of the genetic diversity that is the basis for a species' adaptability and continued evolution, preserving the phenotype but not its full genotype.

The role of captive breeding as a conservation tool is discussed by Foose (1983) and Foose & Flesness (1991). George (1990a) sets out the case for its use in the Australasian context. George (1990b) surveys the captive breeding history of Australasian marsupials and monotremes and discuss the role of zoos in captive breeding for conservation purposes. Zoos have limited resources but considerable expertise in captive breeding, and many are establishing collaborative captive breeding programs in cooperation with conservation agencies. Zoos also cooperate with each other internationally through regional breeding programs. The Species Management Program of the Australasian zoos interacts with parallel programs in the U.S.A., U.K., and Europe based on regional and international studbooks (Baker & George 1988). Similar programs are being developed in other regions. Leadbeater's possum (*Gymnobelideus leadbeateri*), brush-tailed bettong, and various New Guinea tree-kangaroos are already being maintained and bred in overseas zoos under these studbook-based breeding programs. There is considerable potential to expand these links to augment the limited accommodation available in Australasian zoos.

The IUCN policy on captive breeding stresses the value of self-sustaining captive populations as support for wild populations. Captive populations can be used as the source of animals for translocation. To be of conservation value captive populations need to be closely managed as part of the meta-population and preferably, interactively with other sub-populations. This is being done in Western Australia with the numbat and in Victoria with the eastern barred bandicoot and Leadbeater's possum. In South Australia, captive bred brush-tailed bettongs were used for the island introductions. Captive-bred animals usually require extensive acclimatization, monitoring, and support for a period after release to aid successful establishment.

Leadbeater's possum, *Gymnobelideus leadbeateri* (photo by J.H. Seebeck, Victorian Department of Conservation and Environment).

Population Viability Analysis

Management of a population requires adequate planning: identifying the problem, establishing goals, assessing the probability of success, providing resources, and monitoring results. In the planning process, genetic and demographic management are as important as the control of other threatening processes. Many, if not most, historical attempts to establish safe populations of threatened species in refuges such as island reserves, have failed, because of inadequate planning and monitoring. Planning and implementing recovery programs can be very costly and limitations on resources usually require priorities to be established. In a case study Maguire et al. (1986) describe the use of "Decision Analysis" as an analytical tool to assess the risk of extinction, the risks and benefits of alternative management strategies, and the justification of costs.

Gilpin and Soulé (1986) introduced the term Population Vulnerability Analysis (PVA) to cover the process of determining the Minimum Viable Population (MVP) size needed to ensure a high probability of persistence for any given threatened population. Franklin (1980) made a preliminary suggestion that an Effective Population Size of 500 would be the minimum necessary to ensure long-term population viability, based on long-term maintenance of a balance of genetic diversity. (In most natural populations the genetically effective size [Ne] is much less than the census size [N]—Lande & Barrowclough (1987) discuss the mathematics involved). Shaffer (1981) discussed the MVP concept, pointing out that stochastic (random, unpredictable) events mean that much larger population sizes are needed to ensure a population's persistence. Gilpin and Soulé included deterministic processes, such as habitat fragmentation, in their PVA procedure, confirming that there is no "magic number" that will ensure the persistence of all species.

Under the auspices of the Captive Breeding Specialist Group, PVA has come into extensive use (as Population Viability Analysis) as a tool for the recovery of several high-profile threatened species, for example, the Puerto Rican parrot (Lacy, Flesness & Seal 1989), Florida panther (Seal & Lacy 1989), and Javan rhinoceros (Seal & Foose 1989). These PVA's include management recommendations such as Species Survival Plans.

Eight case studies in Victoria were used to illustrate the concepts and tools involved in PVA at a conference devoted to small population management and conservation at Melbourne Zoo in September, 1989. Decision analysis was used to compare alternative management strategies for the eastern barred bandicoot (Maguire et al. 1990) and computer simulation software (SIMPOP) was introduced as a tool for modelling the fate of threatened populations (Lacy & Clark 1990). A later version of this software, VORTEX, was demonstrated at a workshop at the Arthur Rylah Institute in May, 1990, where it was used by departmental and zoo biologists to test management options for several threatened taxa in Victoria (Clark et al. 1991).

VORTEX, written by Bob Lacy and run on MS-DOS microcomputers, simulates demographic and genetic events in the history of a chosen small population. VORTEX models population processes as discrete sequential events, using a pseudorandom number generator to determine outcomes as probabilities. The operator enters life history data and chooses appropriate levels of environmental variation, likelihood of catastrophes, severity of inbreeding, number of simulations to be run and the time span. The program tracks the fate of each simulated population and produces summary statistics on the probability of extinction, mean time to extinction, mean size of surviving populations and levels of remaining genetic variation. The effect of supplementing or harvesting the population at varying rates can be tested by re-running the simulation with varying values for these parameters (Clark et al. 1991).

VORTEX, originally based on algorithms developed by James Grier, is being continuously updated in response to developments in conservation genetics. An improved version, capable of modelling multiple populations, as in a meta-population, has been developed (Lacy, pers. comm), and is available from the CBSG office.

6. Action Plan for Australasian Marsupials and Monotremes

Assessment System Used in the Action Plan

The species conservation assessment used herein draws heavily upon the criteria used in two recent reviews by Burbidge and McKenzie (1989) and Strahan (unpublished ms.). It maintains the distribution and density scales used by Strahan, further refines the percent declined scale used by Burbidge and McKenzie, and, for Australian species, indicates whether the species is in the "critical weight range" developed by the latter authors. This assessment system has also been applied to New Guinea species.

The "Endangered" and "Vulnerable" assignments given in this plan were determined according to IUCN definitions as amended in the "Australian National Strategy for the Conservation of Species and Habitats Threatened with Extinction" (1989):

Endangered: Species in danger of extinction and whose survival is unlikely if the causal factors continue operating. Included are those species whose numbers have been reduced to a critical level or whose habitats have been so drastically reduced that they are deemed to be in immediate danger of extinction. Also included are species that are possibly extinct but have definitely been seen in the wild in the past 50 years and have not been the subject of recent thorough searches.

Vulnerable: Species believed likely to move into the 'Endangered" category in the near future if the causal factors continue operating. Included are species of which most or all the populations are decreasing because of overexploitation, extensive destruction of habitat or other environmental disturbance; species with populations that have been seriously depleted and whose ultimate security has not yet been assured; and species with populations that are still abundant but are under threat from severe adverse factors throughout their range. Also included are species with low or localised populations or dependent upon limited habitat that would be vulnerable to new threatening processes.

The other categories are as follows:

Potentially Vulnerable: has been used in this plan to convey the concerns and approaches taken by Ride and Wilson (1982), Strahan (unpublished ms), and Burton, Kennedy and Fry (1986) for lesser conservation ratings:

Ride and Wilson: "...the state of a population which is thought to be secure, but which requires vigilance and protection to prevent predictably deleterious processes from commencing to operate."

Strahan: "Of satisfactory distribution and/or abundance but faced with present or foreseeable pressures which could lead the species to become endangered."

Burton, Kennedy and Fry: "This category is a type of "catch-all" for species which would appear to need some form of monitoring due to a "general vulnerability" but which cannot be slotted into existing IUCN categories."

Apparently Stable: also coincides with "Probably Secure" as used by Strahan: "Apparently as below but insufficient data."

Stable: coincides with the prognosis of "Secure" used by Strahan: "No existing or foreseeable threat to continuance as abundant or common species."

Red-tailed phascogale, *Phascogale calura* (photo by J.H. Seebeck, Victorian Department of Conservation and Environment).

Probably Extinct: as per Ride and Wilson. Based on the IUCN definition of "extinct", "species not definitely located in the wild during the past 50 years," but with a plea for sparing use given Australia's record of rediscoveries.

Strategies for the Conservation of Australasian Marsupials and Monotremes

Recovery outlines have been developed for the 19 endangered and eight vulnerable Australian marsupials. Each of the Australian outlines provides a brief assessment of the distribution and status of the species and an overview of the factors considered responsible for its decline. The research and management objectives relevant to the species are then discussed followed by a summary of the project costs, in 1991 Australian dollars, for the most urgent actions required. Recovery outlines for Australian species are given in Appendix II.

Insufficient information is available to prepare similar outlines for marsupials and monotremes under threat in Indonesia and Papua New Guinea. Basic information about those species thought to be endangered or vulnerable, however, is given in Table 22.

Table 23 provides a list of all Australasian marsupials and monotremes currently on the Appendices to CITES and/or listed in IUCN's Red List of Threatened Animals (1990). The table also gives the new proposed categories for IUCN listing, though recommendations for CITES listings and delistings will form part of subsequent AMMSG activities.

Australasian Species Requiring Monitoring: A Reserve List

Some Australasian species, identified as potentially vulnerable in this plan, are in need of regular monitoring, but not fully developed recovery plans. This may be for reasons of restricted range, insufficient ecological information, insecure habitat or indeterminate status. These species must be subjected to periodical monitoring programs throughout their ranges. They represent a second level of monitoring, less intensive than that required for species for which recovery outlines are presented. Potentially vulnerable species in Australasia are listed in Table 24.

The Conservation Status of Australasian Marsupials and Monotremes

The coding for columns 1-5 in the status table below (Table 21) is as follows:

Column 1—Status as derived for this plan:

ST—Stable
AS—Apparently Stable
PV—Potentially Vulnerable
V—Vulnerable
E—Endangered
PE—Probably Extinct

Column 2—% Declined (geographic range, since European settlement):

INCR—Increased
<10—No change or declined < 10%
10-50—Declined between 10% & 50%
50-90—Declined between 50% & 90%
>90—Declined > 90%
100—Extinct

Column 3—Critical Weight Range (application to Australian species only):

"Y" indicates that the species is within the critical weight range (CWR) of Burbidge and McKenzie (1989), i.e. it has a mean adult body weight of between 35 g and 5,500 g.

Column 4—Captive breeding classification under the Australasian Species Management Program of the major zoos:

1: Managed—Conservation Programs. Taxa managed regionally by zoos in collaboration with wildlife management agencies (where return to the wild of captive-bred animals is an integral part of management plans)

2: Managed—Regional Zoo Programs. Taxa managed cooperatively on a regional basis (because of potential conservation value, high profile display value, and restricted availability, i.e. rare in the wild or subject to quarantine or other controls)

3: Monitored—Taxa monitored regionally via studbook registration (managed as sub-populations of other regional programs, international studbook species with small regional populations, and taxa managed via local in-house programs)

4: Censused—Taxa for which there is no genetic concern (common in captivity, or readily available from the wild, requiring monitoring only by annual census)

The assignment of management categories to species within the Australian zoo collection is subject to ongoing review to include updated information on conservation status. An asterisk used in conjunction with a category definition indicates a flag for reclassification.

Column 5—Marsupial and monotreme regional occurrence:

I—Irian Jaya/Indonesian Islands
P—Papua New Guinea
S—Solomon Islands
A—Australia

The distribution indicated for Irian Jaya and Papua New Guinea (New Guinea mainland) follows Flannery (1990). Where only "I" or "P" is shown, this simply reflects recorded specimens to date. Further work may well show the species to occur throughout New Guinea.

Table 21. The conservation status of Australasian marsupials and monotremes.

	Status 1	% Declined 2	Critical Weight 3	Captive Breeding 4	Country 5
Prototheria					
Order Monotremata					
Family Tachyglossidae					
Tachyglossus aculeatus (Shaw 1792), Short-beaked echidna	ST	<10	Y	4	IPA
Zaglossus bruijni (Peters & Doria 1876), Long-beaked echidna	V	50-90	-	4	IP
Family Ornithorhynchidae					
Ornithorhynchus anatinus (Shaw 1799), Platypus	PV[1]	<10	Y	4	A
Marsupialia					
Order Dasyuromorphia					
Family Thylacinidae					
Thylacinus cynocephalus (Harris 1808), Thylacine	PE	100	-	-	A
Family Myrmecobiidae					
Myrmecobius fasciatus Waterhouse 1836, Numbat	E	>90	Y	1	A
Family Dasyuridae					
Antechinomys laniger (Gould 1856), Kultarr	PV	<10	Y	3	A
Antechinus bellus (Thomas 1904), Fawn antechinus	ST	<10	Y	-	A
Antechinus flavipes (Waterhouse 1838), Yellow-footed antechinus	ST	<10	Y	-	A
Antechinus godmani (Thomas 1923), Atherton antechinus	PV	<10	Y	3*	A
Antechinus leo Van Dyck 1980, Cinnamon antechinus	ST	<10	Y	-	A
Antechinus minimus (Geoffroy 1803), Swamp antechinus	AS	<10	Y	-	A
Antechinus stuartii Macleay 1841, Brown antechinus	ST	<10	Y	-	A
Antechinus swainsonii (Waterhouse 1840), Dusky antechinus	AS	<10	Y	4	A
"Antechinus" melanurus (Thomas 1899), Black-tailed antechinus	AS	<10	-	-	IP
"Antechinus" habbemma Tate & Archbold 1941, Lesser antechinus	AS	<10	-	-	IP
"Antechinus" naso (Jentink 1911), Long-nosed antechinus	AS	<10	-	-	IP
Dasycercus (Dasyuroides) byrnei (Spencer 1896), Kowari	E	50-90	Y	4	A
Dasycercus cristicauda (Krefft 1867), Mulgara	V	50-90	Y	3*	A
Dasyurus albopunctatus Schlegel 1880, New Guinea quoll	AS	<10	-	-	IP
Dasyurus geoffroii Gould 1841, Chuditch/Western quoll	E	>90	Y	3*	A
Dasyurus hallucatus Gould 1842, Northern quoll	AS	10-50	Y	4	A
Dasyurus maculatus (Kerr 1792), Tiger quoll	PV	50-90	Y	3	A
Dasyurus spartacus Van Dyck 1988, Bronze quoll	PV	?	-	-	P
Dasyurus viverrinus (Shaw 1800), Eastern quoll	PV[2]	50-90	Y	4	A
Murexia longicaudata (Schlegel 1866), Short-furred dasyure	AS	<10	-	-	IP
Murexia rothschildi Tate 1938, Broad-striped dasyure	AS	<10	-	-	P
Myoictis melas (Muller 1840), Short-furred dasyure	AS	<10	-	-	IP
Neophascogale lorentzi (Jentink 1911), Speckled dasyure	AS	<10	-	-	IP
Ningaui ridei Archer 1975, Wongai ningaui	ST	<10	-	-	A
Ningaui timealeyi Archer 1975, Pilbara ningaui	ST	<10	-	-	A
Ningaui yvonneae Kitchener, Stoddart & Henry 1983, Southern ningaui	ST	<10	-	-	A
Parantechinus apicalis (Gray 1842), Dibbler	E	>90	Y	-	A
Parantechinus (Pseudantechinus) bilarni (Johnson 1954), Sandstone antechinus	AS	<10	Y	-	A
Parantechinus (Pseudantechinus) ningbing Kitchener 1988, Ningbing antechinus	ST	<10	-	-	A
Parantechinus (Pseudantechinus) macdonnellensis (Spencer 1895), Fat-tailed antechinus	ST	<10	-	-	A
Parantechinus (Pseudantechinus) mimulus (Thomas 1906), Carpentarian antechinus	PV	?	-	-	A
Parantechinus (Pseudantechinus) woolleyae Kitchener & Caputi 1988, Woolley's antechinus	AS	<10	-	-	A
Parantechinus (Dasykaluta) rosamondae (Ride 1964), Little red antechinus	PV	<10	Y	-	A
Phascogale calura Gould 1884, Red-tailed phascogale	E	>90	Y	-	A
Phascogale tapoatafa (Meyer 1793), Brush-tailed phascogale	PV	10-50	Y	3*	A
Phascolosorex doriae (Thomas 1896), Red-bellied dasyure	AS	<10	-	-	I
Phascolosorex dorsalis (Peters & Doria 1876), Narrow-striped dasyure	AS	<10	-	-	IP
Planigale gilesi Aitken 1972, Paucident planigale	ST	<10	-	-	A
Planigale ingrami (Thomas 1906), Long-tailed planigale	ST	<10	-	-	A
Planigale maculata (Gould 1851), Common planigale	ST	<10	-	-	A
Planigale novaeguineae Tate & Archbold 1941, New Guinea planigale	AS	<10	-	-	P
Planigale tenuirostris Troughton 1928, Narrow-nosed planigale	ST	<10	-	-	A
Sarcophilus harrisii (Boitard 1841), Tasmanian devil	ST	INCR?	-	3*	A

Table 21. Continued

	Status 1	% Declined 2	Critical Weight 3	Captive Breeding 4	Country 5
Sminthopsis aitkeni Kitchener, Stoddart & Henry 1984, Kangaroo Island dunnart	PV	<10?	-	-	A
Sminthopsis archeri van Dyck 1986, Chestnut dunnart (Australian population)	AS	<10	-	-	A
Sminthopsis archeri van Dyck 1986, Chestnut dunnart (New Guinea population)	PV	?	-	-	P
Sminthopsis butleri Archer 1979, Carpentarian dunnart	PV	10-50	-	-	A
Sminthopsis crassicaudata (Gould 1844), Fat-tailed dunnart	ST	<10	-	4	A
Sminthopsis dolichura Kitchener, Stoddart & Henry 1984, Little long-tailed dunnart	ST	<10	-	-	A
Sminthopsis douglasi Archer 1979, Julia Creek dunnart	E	<10?	Y	-	A
Sminthopsis fuliginosus (Gould 1852), Dunnart, no common name	?	?	-	-	A
Sminthopsis gilberti Kitchener, Stoddart & Henry 1984, Gilbert's dunnart	AS	<10	-	-	A
Sminthopsis granulipes Troughton 1932, White-tailed dunnart	AS	<10	-	-	A
Sminthopsis griseoventer Kitchener, Stoddart & Henry 1984, Grey-bellied dunnart	ST	<10	-	-	A
Sminthopsis hirtipes Thomas 1898, Hariy-footed dunnart	ST	<10	-	-	A
Sminthopsis leucopus (Gray 1842), White-footed dunnart	AS	<10	-	-	A
Sminthopsis longicaudata Spencer 1909, Long-tailed dunnart	AS	<10	-	-	A
Sminthopsis macroura (Gold 1845), Stripe-faced dunnart	ST	<10	-	4	A
Sminthopsis murina (Waterhouse 1838), Common dunnart	ST	<10	-	-	A
Sminthopsis ooldea Troughton 1965, Ooldea dunnart	ST	<10	-	-	A
Sminthopsis psammophila Spencer 1895, Sandhill dunnart	V	?	-	-	A
Sminthopsis virginiae (Tarragon 1847), Red-cheeked dunnart	PV	<10	Y	-	PA
Sminthopsis youngsoni McKenzie & Archer 1982, Lesser hairy-footed dunnart	ST	<10	-	-	P
Order Peramelemorphia					
Family Peramelidae					
Chaeropus ecaudatus (Ogilby 1838), Pig-footed bandicoot	PE	100	-	-	A
Isoodon auratus (Ramsay 1887), Golden bandicoot	E	<90	Y	-	A
Isoodon macrourus (Gould 1842), Northern brown bandicoot	ST	10-50	Y	4	PA
Isoodon obesulus (Shaw 1797), Southern brown bandicoot	PV	50-90	Y	4	A
Perameles bougainville Quoy & Gaimard 1824, Western barred bandicoot	E	<90	Y	-	A
Perameles eremiana Spencer 1897, Desert bandicoot	PE	100	-	-	A
Perameles gunnii Gray 1838, Eastern barred bandicoot	V[4]	<90	Y	3*	A
Perameles nasuta Geoffroy 1804, Long-nosed bandicoot	ST	<10	Y	4	A
Macrotis lagotis (Reid 1837), Bilby/Dalgyte	V[5]	50-90	Y	2*	A
Macrotis leucura (Thomas 1887), Lesser bilby	PE	100	Y	-	A
Family Peroryctidae					
Echymipera davidi Flannery (1990), Kiriwina bandicoot	PV	<10	-	-	P
Echymipera clara Stein 1932, Clara or Dimorphic bandicoot	PV	<10	-	-	IP
Echymipera kalubu (Lesson 1828), Kalubu bandicoot	AS	<10	-	-	IP
Echymipera rufescens (Peters & Doria 1875), Rufous spiny bandicoot (Australian population)	PV	<10	Y	-	A
Echymipera rufescens (Peters & Doria 1875), Rufous spiny bandicoot (New Guinea population)	AS	<10	-	-	IP
Echymipera echinista Menzies 1990, Fly River bandicoot	PV	<10?	-	-	P
Peroryctes braodbenti (Ramsay 1879), Broadbent/Giant bandicoot	E	50-90	-	-	P
Peroryctes rafrayanus (Milne-Edwards 1878), Raffray bandicoot	AS	<10	-	-	IP
Microperoryctes longicauda (Peters & Doria 1876), Striped bandicoot	AS	<10	-	-	IP
Microperoryctes murina Stein 1932, Mouse bandicoot	PV	<10	-	-	I
Microperoryctes papuensis (Laurie 1952), Papuan bandicoot	AS	<10	-	-	P
Rhynchomeles prattorum Thomas 1920, Seram bandicoot	V	?	-	-	I
Order Notoryctemorphia					
Family Notoryctidae					
Notorcytes typhlops (Stirling 1889), Marsupial mole	PV	<10	Y	4	A
Order Diprotodontia					
Sub-order Vombatiformes					
Family Phascolarctidae					
Phascolarctos cinereus (Goldfuss 1817), Koala	PV[6]	50-90	-	3	A
Family Vombatidae					
Lasiorhinus krefftii (Owen 1872), Northern hairy-nosed wombat	E	>90	-	-	A

Table 21. Continued.

	Status 1	% Declined 2	Critical Weight 3	Captive Breeding 4	Country 5
Lasiorhinus latifrons (Owen 1845), Southern hairy-nosed wombat	AS	10-50	-	4	A
Vombatus ursinus (Shaw 1800), Common wombat	AS	10-50	-	4	A
Sub-order Phalangerida					
Family Phalangeridae					
Ailurops ursinus (Temminck 1824), Large Celebes cuscus	PV	?	-	-	I
Phalanger carmelitae Thomas 1896, Mountain cuscus	AS	<10	-	-	IP
Phalanger g. gymnotis (Peters & Doria 1875), Aru Islands ground cuscus	PV	?	-	-	I
Phalanger g. leucippus Thomas 1898, New Guinea ground cuscus	AS	<10	-	4	IP
Phalanger lullulae Thomas 1896, Woodlark Island cuscus	E	10-50?	-	-	P
Phalanger matanim Flannery 1987, Telefomin cuscus	V	?	-	-	P
Phalanger orientalis (Pallas 1766), Grey cuscus	ST	<10	-	-	IPSA
Phalanger ornatus Gray 1860, Moluccan cuscus	PV	?	-	-	I
Phalanger pelengensis Tate 1945, Peleng Island cuscus	PV	?	-	-	I
Phalanger rothschildi Thomas 1898, Obi Island cuscus	PV	?	-	-	I
Phalanger sericeus Thomas 1907, Silky cuscus	AS	<10	-	-	IP
Phalanger vestitus Milne-Edwards 1877, Stein's cuscus	PV	?	-	-	IP
Spilocuscus kraemeri (Schwarz 1910), Manus Island spotted cuscus	AS	<10	-	-	P
Spilocuscus maculatus (Desmarest 1818), Spotted cuscus	AS	<10	-	4	IPA
Spilocuscus papuensis (Desmarest 1922), Waigeo spotted cuscus	PV	?	-	-	I
Spilocuscus rufoniger (Zimara 1937), Black-spotted cuscus	V	50-90?	-	-	IP
Strigocuscus celebensis (Gray 1858), Little Celebes cuscus	PV	?	-	-	I
Trichosurus arnhemensis Collett 1897, Northern brushtail possum	PV	10-50	Y	4	A
Trichosurus caninus (Ogilby 1836), Mountain brushtail possum	PV	<10	Y	4	A
Trichosurus vulpecula (Kerr 1792), Common brushtail possum	PV[7]	10-50	Y	4	A
Wyulda squamicaudata Alexander 1919, Scaly-tailed possum	AS	<10	Y	-	A
Family Potoroidae					
Aepyprymnus rufescens (Gray 1837), Rufous bettong	PV	10-50	Y	4	A
Bettongia gaimardi (Desmarest 1822), Tasmanian bettong	PV	50-90	Y	-	A
Bettongia lesueur (Quoy & Gaimard 1824), Boodie/Burrowing bettong	E	>90	Y	-	A
Bettongia penicillata (Gray 1837), Woylie/Brush-tailed bettong	E	>90	Y	-	A
Bettongia tropica Wakefield 1967, Northern bettong	E	?	Y	-	A
Caloprymnus campestris (Gould 1943), Desert rat-kangaroo	PE	100	Y	-	A
Hypsiprymnodon Moschatus Ramsay 1876, Musky rat-kangaroo	PV	10-50	Y	4	A
Potorous longipes Seebeck & Johnston 1980, Long-footed potoroo	E	?	Y	1	A
Potorous platyops (Gould 1844), Broad-faced potoroo	PE	100	Y	-	A
Potorous tridactylus (Kerr 1792), Long-nosed potoroo	PV[8]	10-50	Y	4	A
Family Macropodidae					
Dendrolagus bennettianus De Vis 1887, Bennett's tree-kangaroo	AS	?	-	-	A
Dendrolagus lumholtzi Collett 1884, Lumholtz's tree-kangaroo	AS	?	-	-	A
Dendrolagus dorianus Ramsay 1883, Doria's tree-kangaroo	V	10-50	-	-	IP
Dendrolagus goodfellowi Thomas 1908, Goodfellow's tree-kangaroo	V	50-90?	-	3	IP
Dendrolagus matschiei Forster & Rothschild 1907, Matschie's tree-kangaroo	PV	10-50	-	3	P
Dendrolagus spadix Troughton & Le Souef 1936, Lowland tree-kangaroo	PV	<10?	-	-	P
Dendrolagus scottae Flannery & Seri (1990), Scott's tree-kangaroo	E	50-90?	-	-	P
Dendrolagus ursinus Muller 1840, White-throated tree-kangaroo	PV	?	-	-	I
Dendrolagus inustus Muller 1840, Grizzled tree-kangaroo	PV	?	-	-	IP
Dorcopsis atrata Van Deusen 1955, Black dorcopsis	PV	<10?	-	-	P
Dorcopsis hageni Jeller 1897, White-striped dorcopsis	AS	<10	-	-	IP
Dorcopsis luctuosa (D'Albertis 1874), Grey dorcopsis	AS	<10	-	3*	IP
Dorcopsis muelleri (Schlegel 1866), Brown dorcopsis	AS	<10?	-	-	IP
Dorcopsulus macleayi (Mikluho-Maclay 1885), Macleay's dorcopsis	V	50-90?	-	-	P
Dorcopsulus vanheurni (Thomas 1892), Little dorcopsis	AS	<10	-	-	IP
Lagorchestes conspicillatus Gould 1842, Spectacled hare-wallaby	PV	10-50	Y	4	A
Lagorchestes hirsutus Gould 1844, Mala/Rufous hare-wallaby	E	>90	Y	3*	A
Lagorchestes leporides (Gould 1841), Eastern hare-wallaby	PE	100	Y	-	A
Lagorchestes asomatus Finlayson 1943, Central hare-wallaby	PE	100	Y	-	A
Lagostrophus fasciatus (Peron & Lesueur 1807), Banded hare-wallaby	E	>90	Y	-	A
Macropus agilis (Gould 1842), Agile wallaby	ST	<10	-	4	IPA
Macropus antilopinus (Gould 1842), Antilopine wallaroo	ST	<10	-	4	A
Macropus bernardus Rothschild 1904, Black wallaroo	ST	<10	-	-	A

Table 21. Continued.

	Status 1	% Declined 2	Critical Weight 3	Captive Breeding 4	Country 5
Macropus dorsalis (Gray 1837), Black-striped wallaby	ST	10-50	-	4	A
Macropus eugenii (Desmarest 1817), Tammar wallaby	V	>90	Y	4	A
Macropus fuliginosus (Desmarest 1817), Western grey kangaroo	ST	<10	-	4	A
Macropus giganteus Shaw 1790, Eastern grey kangaroo	ST[9]	INCR	-	4	A
Macropus greyi Waterhouse 1845, Toolache wallaby	PE	100	-	-	A
Macropus irma (Jourdan 1837), Western brush wallaby	PV	<10	-	-	A
Macropus parma Waterhouse 1845, Parma wallaby	AS	10-50	Y	4	A
Macropus parryi Bennett 1835, Whiptail wallaby	ST	<10	-	4	A
Macropus robustus Gould 1841, Common wallaroo/Euro	ST[10]	INCR	-	4	A
Macropus rufogriseus (Desmarest 1817), Red-necked wallaby	ST	<10	-	4	A
Macropus rufus (Desmarest 1822), Red kangaroo	ST	INCR	-	4	A
Onychogalea fraenata (Gould 1841), Bridled nailtail wallaby	E	>90	Y	-	A
Onychogalea lunata (Gould 1841), Crescent nailtail wallaby	PE	100	Y	-	A
Onychogalea unguifera (Gould 1841), Northern nailtail wallaby	AS	<10	Y	4	A
Petrogale assimilis Ramsay 1877, Allied rock-wallaby	ST	<10	Y	-	A
Petrogale brachyotis (Gould 1841), Short-eared rock-wallaby	ST	<10	Y	-	A
Petrogale burbidgei Kitchener & Sanson 1978, Monjon/Warabi	ST	<10	Y	-	A
Petrogale concinna Gould 1842, Nabarlek	PV	<10	Y	-	A
Petrogale godmani Thomas 1923, Godman's rock-wallaby	PV	<10	Y	-	A
Petrogale inornata Gould 1842, Unadorned rock-wallaby	ST	<10	Y	-	A
Petrogale lateralis Gould 1842, Black-footed rock-wallaby	V	>90	Y	-	A
Petrogale penicillata (Gray 1825), Brush-tailed rock-wallaby	V	50-90	-	4	A
Petrogale persephone Maynes 1982, Prosperpine rock-wallaby	E	?	-	-	A
Petrogale rothschildi Thomas 1904, Rothschild's rock-wallaby	AS	<10	Y	-	A
Petrogale xanthopus Gray 1855, Yellow-footed rock-wallaby	PV[11]	50-90	-	2*	A
Setonix brachyurus (Quoy & Gaimard 1830), Quokka	PV	50-90	Y	4	A
Thylogale billardierii (Desmarest 1822), Tasmanian pademelon	ST	10-50	Y	4	A
Thylogale brunii (Schreber 1778), Brown/Dusky pademelon	ST	<10	-	-	IP
Thylogale stigmatica (Gould 1860), Red-legged pademelon	AS	<10	Y	4	PA
Thylogale thetis (Lesson 1827), Red-necked pademelon	AS	<10	Y	4	A
Wallabia bicolor (Desmarest 1804), Swamp wallaby	ST	<10	-	4	A
Family Burramyidae					
Burramys parvus Broom 1896, Mountain pygmy-possum	E	10-50	Y	3	A
Cercartetus caudatus (Milne-Edwards 1877), Long-tailed pygmy-possum	ST	<10	-	-	IPA
Cercartetus concinnus (Gould 1845), Western pygmy-possum	ST	<10	-	-	A
Cercartetus lepidus (Thomas 1888), Little pygmy-possum	PV	?	-	-	A
Cercartetus nanus (Desmarest 1818), Eastern pygmy-possum	AS	<10	-	4	A
Family Pseudocheiridae					
Hemibelideus lemuroides (Collett 1884), Lemuroid ringtail possum	AS	<10	Y	-	A
Petauroides volans (Kerr 1792), Greater glider	PV	<10	Y	4	A
Petropseudes dahli (Collett 1895), Rock ringtail possum	ST	?	Y	4	A
Pseudocheirus canescens (Waterhouse 1846), Lowland/Hoary ringtail	AS	?	-	-	IP
Pseudocheirus caroli Thomas 1921, Weyland ringtail	PV	?	-	-	I
Pseudocheirus forbesi Thomas 1897, Painted ringtail	AS	<10	-	-	IP
Pseudocheirus herbertensis (Collett 1884), Herbert River ringtail possum	AS	?	Y	-	A
Pseudocheirus cinercus Tate 1945, Daintree River ringtail possum	AS	?	Y	-	A
Pseudocheirus mayeri Rothschild & Dollman 1932, Pygmy ringtail	AS	<10	-	-	IP
Pseudocheirus occidentalis Thomas 1888, Western ringtail	V	50-90	Y	-	A
Pseudocheirus peregrinus (Boddaert 1785), Common ringtail possum	AS	<10	Y	4	A
Pseudocheirus schlegeli Jentink 1884, Arfak ringtail	PV	?	-	-	I
Pseudochirops albertisi (Peters 1874), D'Albertis ringtail	AS?	?	-	-	IP
Pseudochirops archeri (Collett 1884), Green ringtail possum	AS	?	Y	-	A
Pseudochirops corinnae (Thomas 1897), Golden/Plush-coated ringtail	AS	<10	-	-	IP
Pseudochirops cupreus (Thomas 1897), Copper ringtail	AS	<10	-	-	IP
Family Petauridae					
Dactylopsila trivirgata Gray 1858, (Common) Striped possum	ST	<10	Y	4	IPA
Dactylopsila megalura Rothschild & Dollman 1932, Great-tailed triok/Striped possum	V	?	-	-	IP
Dactylopsila tatei Laurie 1952, Fergusson Island striped possum	PV	?	-	-	P
Dactylopsila palpator Milne-Edwards 1888, Long-fingered triok	AS	<10	-	-	IP
Gymnobelideus leadbeateri McCoy 1867, Leadbeater's possum	E	10-50	Y	1	A
Petaurus abidi Zeigler 1981, Torricelli/Northern glider	V	?	-	-	P
Petaurus australis Shaw 1791, Yellow-bellied glider	PV	<10	Y	3*	A

Table 21. Continued

	Status 1	% Declined 2	Critical Weight 3	Captive Breeding 4	Country 5
Petaurus breviceps Waterhouse 1839, Sugar glider	AS	<10	Y	4	IPA
Petaurus gracilis (de Vis 1883), Mahogany glider	?[12]	?	?	-	A
Petaurus norfolcensis (Kerr 1792), Squirrel glider	PV	10-50	Y	4	A
Family Tarsipedidae					
Tarsipes rostratus Gervais & Verreaux 1842, Honey-possum	PV	10-50	-	4	A
Family Acrobatidae					
Acrobates pygmaeus (Shaw 1794), Feathertail glider	AS	<10	-	4	A
Distoechurus pennatus (Peters 1874), Feather-tailed possum	AS	<10	-	-	IP

Footnotes*

[1]Platypus—"Apparently Stable" in Tasmania, "Vulnerable" in South Australia.
[2]Eastern quoll—"Probably Extinct" on the mainland.
[3]*Sminthopsis fuliginosus*—More information required. Western Australia.
[4]Eastern barred bandicoot—"Endangered" in Victoria.
[5]Bilby—"Endangered" in Queensland.
[6]Koala—"Apparently Stable" in Victoria, "Vulnerable" in New South Wales.
[7]Common brushtail possum—"Endangered" in Northern Territory.
[8]Gilbert's potoroo (*Potorous tridactylus gilberti*)—"Probably Extinct" in Western Australia.
[9]Eastern grey kangaroo (*Macropus giganteus tasmaniensis*)—"Potentially Vulnerable" in Tasmania.
[10]Common wallaroo/Euro (*Macropus robustus isabellinus*)—"Vulnerable" on Barrow Island, Western Australia.
[11]Yellow-footed rock-wallaby—"Endangered" in New South Wales.
[12]Mahogany glider—Recently rediscovered species in Queensland. More information required, very confined range, may be endangered.

*The species status determinations in this Action Plan have been developed from a national perspective. Conservation status at a state or regional level may be worse, or, alternatively, more stable. The footnotes above do not represent an exhaustive review of the status of marsupials and monotremes at the state level, but merely instances where the AMMSG felt the information was important to note.

Sandhill dunnart, *Sminthopsis psammophila* (photo by P. Canty, South Australian National Parks and Wildlife Service).

Table 22. Endangered and vulnerable species of monotremes and marsupials in Indonesia and Papua New Guinea.

Key to table 22:
1. Reason for listing
2. Distribution
3. Habitat type
4. Recommended conservation action
5. Responsible Wildlife Authority/ies

PHPA: Directorate General of Forest Protection and Nature Conservation, Indonesia
PNGDE&C: Papua New Guinea Department of Environment and Conservation

***Zaglossus bruijni*, Long-beaked echidna**
Vulnerable
1. Decline due to traditional hunting, highly-prized game species
2. Northwest cordillera of Irian Jaya, central cordillera and Huon Peninsula, Papua New Guinea
3. Mid-mountain and lower montane rain forests
4. Reserve and protect known populations, public education, investigate captive breeding
5. PHPA, PNGDE&C

***Peroryctes broadbenti*, Giant bandicoot**
Endangered
1. Very rare, restricted range, game species
2. Southeast Papua New Guinea
3. Lowland rainforest, possibly also lowland hill forest
4. Survey to establish range, status, and ecology
5. PNGDE&C

***Rhynchomeles prattorum*, Seram bandicoot**
Vulnerable
1. Very rare, restricted range
2. Seram Island, Indonesia
3. Mountain forests (?)
4. Survey to establish range, status, and ecology
5. PHPA

***Phalanger lullulae*, Woodlark Island cuscus**
Endangered
1. Restricted range, game species
2. Woodlark Island and Alcester Island, Papua New Guinea
3. Lowland rainforest
4. Survey to establish status, reserve habitat
5. PNGDE&C

***Phalanger matanim*, Telefomin cuscus**
Vulnerable
1. Very rare, restricted altitudinal range and distribution, game species
2. Central cordillera, Papua New Guinea/Irian Jaya border
3. Mountain forests
4. Survey to establish range, status, and ecology
5. PNGDE&C

***Spilocuscus rufoniger*, Black-spotted cuscus**
Vulnerable
1. Rare, limited range, game species
2. Northern Irian Jaya and Papua New Guinea
3. Lowland rainforest
4. Survey to establish status and ecology, protect populations
5. PHPA, PNGDE&C

***Dendrolagus dorianus*, Doria's tree-kangaroo**
Vulnerable
1. Some decline due to hunting pressure, major game species
2. Western and central Irian Jaya through to central and eastern Papua New Guinea
3. Mid-mountain and lower montane forests
4. Reserve habitat, protect populations
5. PHPA, PNGDE&C

***Dendrolagus goodfellowi*, Goodfellow's tree-kangaroo**
Vulnerable
1. Significant range reductions due to overhunting, major game species
2. Border of central Irian Jaya and Papua New Guinea through central and eastern Papua New Guinea
3. Mid-mountain forests
4. Reserve habitat, protect significant populations, extend captive breeding program
5. PHPA, PNGDE&C

***Dendrolagus scottae*, Scott's tree-kangaroo**
Endangered
1. Very rare, very restricted range, game species
2. Northern Papua New Guinea
3. Mossy mountain forests
4. Survey to establish range, status and ecology, start captive breeding
5. PNGDE&C

***Dorcopsulus macleayi*, Macleay's dorcopsis**
Vulnerable
1. Very rare, limited range, game species
2. South-east Papua New Guinea
3. Lowland hill forests
4. Survey to establish range, status, and ecology
5. PNGDE&C

***Dactylopsila megalura*, Great-tailed triok**
Vulnerable
1. Very rare, limited range, game (trophy) species
2. Central cordillera Irian Jaya to Papua New Guinea border
3. Mountain forests
4. Survey to establish status and ecology
5. PHPA, PNGDE&C

***Petaurus abidi*, Northern glider**
Vulnerable
1. Very rare, restricted range
2. Northern Papua New Guinea
3. Mountain forests
4. Survey to establish range, status, and ecology
5. PNGDE&C

Table 23. Current CITES listings and current and proposed IUCN threatened category listings for Australasian marsupials and monotremes (CITES listing/delisting proposals will form part of subsequent AMMSG recommendations).

Scientific Name	Common Name	CITES	IUCN Current	IUCN Proposed
Zaglossus bruijni	Long-beaked echidna	App II	V	V
Thylacinus cynocephalus	Thylacine	App I	Ex	Ex
Myrmecobius fasciatus	Numbat	—	E	E
Dasycercus byrnei	Kowari	—	—	E
Dasycercus cristicauda	Mulgara	—	—	V
Dasyurus geoffroii	Chuditch/Western quoll	—	—	E
Parantechinus apicalis	Dibbler	—	I	E
Phascogale calura	Red-tailed phascogale	—	I	E
Sminthopsis douglasi	Julia Creek dunnart	—	I	E
Sminthopsis longicaudata	Long-tailed dunnart	App I	K	—
Sminthopsis psammophila	Sandhill dunnart	App I	K	V
Chaeropus ecaudatus	Pig-footed bandicoot	App I	Ex	Ex
Isoodon auratus	Golden bandicoot	—	—	E
Perameles bougainville	Western barred bandicoot	App I	R	E
Perameles eremiana	Desert bandicoot	—	Ex	Ex
Perameles gunnii	Eastern barred bandicoot	—	—	V
Macrotis lagotis	Bilby/Dalgyte	App I	E	V
Macrotis leucura	Lesser bilby	App I	Ex	Ex
Echymipera clara	Clara bandicoot	—	R	R
Peroryctes broadbenti	Giant bandicoot	—	—	E
Rhynchomeles prattorum	Seram bandicoot	—	—	V
Lasiorhinus krefftii	Northern hairy-nosed wombat	App I	E	E
Phalanger lullulae	Woodlark Island cuscus	—	E	E
Phalanger matanim	Telefomin cuscus	—	—	V
Phalanger orientalis	Grey cuscus	App II	—	—
Phalanger vestitus	Stein's cuscus	—	R	R
Spilocuscus maculatus	Spotted cuscus	App II	—	—
Spilocuscus rufoniger	Black-spotted cuscus	—	R	V
Bettongia gaimardi	Tasmanian bettong	App I	—	—
Bettongia lesueur	Burrowing bettong/Boodie	App I	R	E
Bettongia penicillata	Brush-tailed bettong/Woylie	App I	E	E
Bettongia tropica	Northern bettong	App I	—	E
Caloprymnus campestris	Desert rat-kangaroo	App I	I	Ex
Potorous longipes	Long-footed potoroo	—	I	E
Potorous platyops	Broad-faced potoroo	—	Ex	Ex
Dendrolagus bennettianus	Bennett's tree-kangaroo	App II	—	—
Dendrolagus lumholtzi	Lumholtz's tree-kangaroo	App II	—	—
Dendrolagus dorianus	Doria's tree-kangaroo	—	V	V
Dendrolagus goodfellowi	Goodfellow's tree-kangaroo	—	V	V
Dendrolagus scottae	Scott's tree-kangaroo	—	—	E
Dendrolagus ursinus	White-throated tree-kangaroo	App II	—	—
Dendrolagus inustus	Grizzled tree-kangaroo	App II	—	—
Dorcopsis atrata	Black dorcopsis	—	R	R
Dorcopsulus macleayi	Macleay's dorcopsis	—	R	V
Lagorchestes hirsutus	Rufous hare-wallaby	App I	R	E
Lagorchestes leporides	Eastern hare-wallaby	—	Ex	Ex
Lagorchestes asomatus	Central hare-wallaby	—	Ex	Ex
Lagostrophus fasciatus	Banded hare-wallaby	App I	R	E
Macropus eugenii	Tammar wallaby	—	—	V
Macropus greyi	Toolache wallaby	—	Ex	Ex
Onychogalea fraenata	Bridled nailtail wallaby	App I	E	E
Onychogalea lunata	Crescent nailtail wallaby	App I	Ex	Ex
Petrogale lateralis	Black-footed rock-wallaby	—	—	V
Petrogale penicillata	Brush-tailed rock-wallaby	—	—	V
Petrogale persephone	Prosperpine rock-wallaby	—	R	E
Burramys parvus	Mountain pygmy-possum	App II	—	E
Pseudocheirus occidentalis	Western ringtail	—	—	V
Dactylopsila megalura	Great-tailed triok	—	—	V
Gymnobelideus leadbeateri	Leadbeater's possum	—	V	E
Petaurus abidi	Northern glider	—	—	V

Key to Table 23:

CITES (Convention on International Trade in Endangered Species of Wild Fauna and Flora) Listings:
I=Appendix I listing (species in which international commercial trade is prohibited)
II=Appendix II listing (species in which international commercial trade is only authorised under permit)

IUCN Category	Abbreviation	Definition
Extinct	Ex	Species not definitely located in the wild during the past 50 years.
Endangered	E	Taxa in danager of extinction and whose survival is unlikely if the causal factors of its decline continue operating.
Vulnerable	V	Taxa believed likely to become endangered in the near future if the causal factors of its decline continue to operate.
Rare	R	Taxa with small world populations that are not at present endangered or vulnerable.
Indeterminate	I	Taxa known to be endangered, vulnerable, or rare, but information is lacking as to which of these categories is appropriate.
Insufficiently Known	K	Taxa that are suspected but not definitely known to belong to any of the above categories due to lack of data.

Note:
This Action Plan makes recommendations only for species which should be listed by IUCN as either endangered or vulnerable, but not for the lesser categories (rare, indeterminate, insufficiently known). The Specialist Group feels that the transference of species in the Action Plan's "Potentially Vulnerable" category into the lesser IUCN categories is problematical. This important exercise will be carried out when IUCN has determined its new conservation status categories in the near future. Three potentially vulnerable marsupials are currently listed by IUCN as "Rare" and the Specialist Group has decided to retain this conservation assessment until the new IUCN categories are agreed upon.

Tammar wallaby, *Macropus eugenii* (photo courtesy of CSIRO-Wildife and Ecology).

Table 24. Potentially vulnerable species of Australasian marsupials and monotremes.

Key to Table 23:
1. Reason for listing
2. Distribution
3. Habitat type
4. Recommended conservation action
5. Responsible wildlife authority/ies

Critical Weight Range (CWR) applies to Australian species only

ACTP&CS: Australian Capital Territory Parks and Conservation Service
CALM: Western Australian Department of Conservation and Land Management
CCNT: Conservation Commission of the Northern Territory
VDCE: Victorian Department of Conservation and Environment
QNPWS: Queensland National Parks and Wildlife Service
NSWNPWS: New South Wales National Parks and Wildlife Service
SANPWS: South Australian National Parks and Wildlife Service
TDPW&H: Tasmanian Department of Parks Wildlife and Heritage
PNGDE&C: Papua New Guinea Department of Environment and Conservation
PHPA: Directorate General of Forest Protection and Nature Conservation, Indonesia

***Ornithorhynchus anatinus*, Platypus** (CWR)
1. Potential vulnerability in aquatic habitat
2. Coastal eastern Australia from north Queensland, south to Victoria and the southeastern tip of South Australia, Tasmania.
3. Suitable bodies of fresh water
4. Compatible management of waterways
5. QNPWS, NSWNPWS, VDCE, SANPWS, TDPW&H, ACTP&CS

***Antechinomys laniger*, Kultarr**
1. Arid and semi-arid species, rare, scattered populations
2. West of the Great Dividing Ranges in New South Wales, southern Queensland and Northern Territory, northern South Australia, and large areas of Western Australia
3. Desert shrublands
4. Population studies, monitoring
5. CALM, CCNT, QNPWS, SANPWS, NSWNPWS

***Antechinus godmani*, Atherton antechinus** (CWR)
1. North Queensland endemic, restricted range and habitat
2. Northeast Queensland
3. High altitude dense forest
4. Ecological, distribution and abundance studies
5. QNPWS

***Dasyurus maculatus*, Tiger quoll** (CWR)
1. Range much reduced on mainland, scarce
2. Northeast coastal Queensland, southeast coastal Queensland, New South Wales, Victoria, Australian Capital Territory and Tasmania
3. Forests
4. Establish status, monitor
5. QNPWS, NSWNPWS, VDCE, TDPW&H, ACTP&CS

***Dasyurus spartacus*, Bronze quoll**
1. Restricted range
2. Southwest coastal Papua New Guinea
3. Trans-Fly savannahs
4. Survey to establish range, status, and ecology
5. PNGDEC

***Dasyurus viverrinus*, Eastern quoll** (CWR)
1. Extinct on mainland, suggestion of some decline in Tasmania
2. Tasmania
3. Open forest, scrub and heath
4. Review Tasmanian status, follow up NSW, Victoria sightings
5. TDPW&H, NSWNPWS, VDCE

***Parantechinus (Pseudantechinus) mimulus*, Carpentarian antechinus**
1. One specimen from mainland, 1905; none since, despite regional surveys
2. Known only from three islands in the Sir Edward Pellew Group, Northern Territory
3. Rocky hills with shrubs and grasses
4. Protection of islands; detailed surveys on the mainland
5. CCNT

***Parantechinus (Dasykaluta) rosamondae*, Little red antechinus**
1. Small range, restricted to arid zone, body weight near CWR
2. Northwest Western Australia
3. Hummock grassland
4. Establish status, study ecology, monitor
5. CALM

***Phascogale tapoatafa*, Brush-tailed phascogale** (CWR)
1. Reduced range, continued habitat destruction
2. Northern coastal Western Australia, Northern Territory and Cape York Peninsula, Queensland, southwest Western Australia, southern/coastal South Australia, Victoria, New South Wales and Queensland, Northern coastal New South Wales
3. Open forest, woodland
4. Appropriate reservation and ecological studies
5. CALM, CCNT, QNPWS, SANPWS, NSWNPWS, VDCE, ACTP&CS

***Sminthopsis aitkeni*, Kangaroo Island dunnart**
1. Extremely restricted range, little known
2. Kangaroo Island, South Australia
3. Open scrub
4. Abundance surveys and ecological studies, studies on effects of predation by feral cats
5. SANPWS

***Sminthopsis archeri*, Chestnut dunnart**
(New Guinea population)
1. Restricted range in New Guinea, no data on population size
2. Southwest coastal Papua New Guinea
3. Trans-Fly savannahs
4. Survey to establish status of New Guinea population, study ecology and monitor
5. PNGDE&C

Table 24. Continued.

***Petrogale godmani*, Godman's rock-wallaby** (CWR)
1. Little known, restricted range, advance of unadorned rock-wallaby
2. Cape York, northeast coastal Queensland
3. Rocky outcrops
4. Ecological studies
5. QNPWS

***Petrogale xanthopus*, Yellow-footed rock-wallaby**
1. Greater than 50% decline in range, continued competition and predation by introduced species
2. Southeast Queensland, northwest New South Wales, and eastern Central South Australia
3. Rock outcrops, semi-arid lands, mulga scrub
4. Protect key colonies, control goats, foxes
5. SANPWS, NSWNPWS, QNPWS

***Setonix brachyurus*, Quokka** (CWR)
1. Declined on mainland
2. Southwest Western Australia
3. Swamps and thickets
4. Monitor, conduct survey of current status
5. CALM

***Cercartetus lepidus*, Little pygmy-possum**
1. Forest fragmentation, inappropriate forest practice
2. Tasmania, Kangaroo Island—South Australia and border between eastern South Australia and northwest Victoria
3. Mallee scrub, dry and wet sclerophyll forests
4. Habitat protection for restricted mainland colonies
5. VDCE, SANPWS, TDPW&H

***Petauroides volans*, Greater glider** (CWR)
1. Continued fragmentation and disturbance of east coast forests
2. Northeast and southeast coastal Queensland, coastal New South Wales, Australian Capital Territory and eastern Victoria
3. Tall eucalypt forests
4. Appropriate forest management, increased forest reservation
5. QNPWS, NSWNPWS, VDCE, ACTP&CS

***Pseudocheirus caroli*, Weyland ringtail**
1. Restricted range, unknown status
2. Western Irian Jaya
3. Rainforest, lowland to mid-mountain
4. Survey to establish range, status, and ecology
5. PHPA

***Pseudocheirus schlegeli*, Arfak ringtail**
1. Very limited range, unknown status, one specimen known
2. Northwestern Irian Jaya-Vogelkon Peninsula
3. Montane (?) forest
4. Survey to establish range, status and ecology
5. PHPA

***Dactylopsila tatei*, Fergusson Island striped possum**
1. Restricted range, unknown status
2. Fergusson Island, southeast Papua New Guinea
3. Lowland rainforest (?)
4. Survey to establish status and ecology
5. PNGDE&C

***Petaurus australis*, Yellow-bellied glider** (CWR)
1. Wet tropics population vulnerable, southern population pressured by continuing logging activities
2. Coastal central and southern Queensland, New South Wales and Victoria, isolated population in northeast Queensland
3. Tall, mature, eucalypt forests, temporate and subtropical
4. Increased reservation, research effects of forestry practices
5. QNPWS, NSWNPWS, VDCE and ACTP&CS

***Petaurus norfolcensis*, Squirrel glider** (CWR)
1. Rare, forest fragmentation
2. Northeast coastal and southern Queensland, central New South Wales, Victoria and Australian Capital Territory, central border between South Australia and Victoria.
3. Dry open sclerophyll forest and woodland
4. Ecological research, forest habitat retention
5. QNPWS, NSWNPWS, VDCE, SANPWS, ACTP&CS

***Tarsipes rostratus*, Honey possum**
1. Habitat loss via clearing for agriculture and *Phytophthora*, possible habitat degradation via inappropriate fire regimes
2. Southwest and coastal Western Australia
3. Kwongan (species-rich sclerophyll heath)
4. Establish appropriate fire regimes and apply in national parks, develop control of *Phytophthora*.
5. CALM

Recommendations

General Recommendations for Indonesia

Recommendation 1
This Action Plan recommends that the government of Indonesia recognise and adopt the entire Irian Jaya protected area system as proposed by WWF, without delay.

Recommendation 2
This Action Plan recommends the adoption by all parties of the guideline boundaries as the first step towards realization of conservation management.

Recommendation 3
This Action Plan recommends the adoption and marking of actual boundaries on the ground appropriate to human needs and conservation requirements, with community support and involvement.

Recommendation 4
This Action Plan recommends active management of the most ecologically important and representative areas of Irian Jaya, with and through community support and involvement.

Dibbler, *Parantechinus apicalis* (photo by M. Morcombe).

Recommendation 5
This Action Plan recommends the management and development of buffer and boundary areas complementary to human needs, to meet the potential for community support and understanding of conservation, and to improve the quality of life for residents near conservation areas.

Recommendation 6
This Action Plan recommends that in developing and dedicating the protected area system in Indonesia, reserves must be sufficiently large to protect self-sustaining and viable populations of all threatened species.

Recommendation 7
This Action Plan recommends that it is essential for the conservation of marsupials in Irian Jaya to include all the major mountain ranges in the protected areas system design.

Recommendation 8
This Action Plan recommends that any Indonesian marsupial or monotreme with a limited geographic or altitudinal range should be considered Potentially Vulnerable. All "Potentially Vulnerable" species identified in this Action Plan should be monitored on as regular a basis as possible.

Specific Recommendations

Recommendation 9
This Action Plan recommends that surveys to establish the status and conservation needs of the endemic marsupials of the Sulawesi and Maluku Provinces be undertaken as a first step towards a conservation program similar to that recommended for Irian Jaya.

Recommendation 10
This Action Plan recommends the reservation and protection of known populations of the long-beaked echidna (*Zaglossus bruijni*), combined with public education programs and the investigation of captive breeding program potential.

Particular management attention should be directed towards the protected areas where the species is known to occur (Lorentz National Park, Arfak Mountains, Tamrau Selatan, and Salawati Utara Strict Nature Reserves) and surveys undertaken to confirm the presence or otherwise of the species in the Jayawijaya Mountains, Tamrau Utara, Weyland Mountains, and Wandamen Mountains Strict Nature Reserves.

Recommendation 11
This Action Plan recommends surveys to establish the range, status and ecology of the Seram bandicoot (*Rhynchomeles prattorum*), and its presence or otherwise in Manusela National Park to assess its conservation needs.

Recommendation 12
This Action Plan recommends surveys to establish the status and ecology of the black-spotted cuscus (*Spilocuscus rufoniger*), and the reservation and protection of all identified populations.

Particular management effort should be directed towards the Cyclops Mountains Strict Nature Reserve, where the species is known to occur, and surveys undertaken to confirm the presence or otherwise of the species in Jayawijaya Mountain, Weyland Mountains, Wandamen and Yapen Tengah Strict Nature Reserves, Rouffaer River Wildlife Sanctuary, and Mamo-Foja National Park.

Recommendation 13
This Action Plan recommends the reservation and protection of all populations of Doria's tree-kangaroo (*Dendrolagus dorianus*).

Particular management efforts should be directed towards the Wandamen Mountains Strict Nature Reserve, where the species is known to occur, and surveys undertaken to confirm the presence or otherwise of the species in Lorentz National park, Jayawijaya Mountains, and Weyland Mountains Strict Nature Reserves.

Recommendation 14
This Action Plan recommends surveys to establish the range, status and ecology of Goodfellow's tree-kangaroo (*Dendrolagus goodfellowi*), the reservation and protection of significant populations of this species, and participation in the captive breeding program.

Surveys should be undertaken to confirm the presence or otherwise of the species in Jayawijaya Mountains Strict Nature Reserve, and the reported sightings in Foja Mountains.

Recommendation 15
This Action Plan recommends surveys to establish the status and ecology of the great-tailed triok (*Dactylopsila megalura*).

Particular management efforts should be directed towards Weyland Mountains Strict Nature Reserve where the species is known to occur, and surveys undertaken to confirm the presence or otherwise of the species in Lorentz National Park and Jayawijaya Strict Nature Reserve.

General Recommendations for Papua New Guinea

Socio-economic and environmental changes are occurring at an increasing pace, yet Papua New Guinea is unable to sustain these impacts effectively through existing institutions. The following recommendations are intended to complement and strengthen the projects outlined in the Papua New Guinea government's National Forest and Conservation Action Plan (NFCAP). Two levels of strategic actions are recommended.

As a component of the NFCAP preparatory stages, the instigation of an "Emergency Strategy" to cater to the needs of identified threatened marsupial and monotreme species:

Recommendation 1
This Action Plan recommends a thorough review and reactivation of existing Wildlife Management Areas (WMA), to complement the rehabilitation program proposed under NFCAP for National Parks and Wildlife Sanctuaries. The process must fully involve all landowners, and place special emphasis in the WMA's management plans on the protection of threatened marsupials and monotremes. It is critical that within threatened species' distributional ranges, areas are set aside exclusively for protection purposes. This will require technical expertise, advice, and financial help in developing appropriate management plans.

Recommendation 2
This Action Plan recommends that special attention be paid, both within WMA management plans and through educational programs, to the increasing problems being faced by marsupials and monotremes through hunting and other pressures near human populations.

Recommendation 3
This Action Plan recommends that wherever the opportunity exists for other compatible use of protected areas, i.e. tourism, that they should be encouraged. It is fundamentally important for the landowners to have some form of cash payment. This should be accomplished in cooperation with the NFCAP "Ecotourism" proposals to be centred around future protected area systems.

Recommendation 4
This Action Plan recommends that all species listed as Endangered or Vulnerable in this Action Plan be considered priority species. These listings should be brought to the attention of regional authorities and the national government and, wherever possible, international assistance should be sought for management programs. These species should be given special consideration in developing management plans for existing and future protected areas.

Recommendation 5
This Action Plan recommends that in cooperation with the Papua New Guinea government (Landowner Awareness and Support

Black-footed rock-wallaby, *Petrogale lateralis* (photo by M. Fleming, Conservation Commission of the Northern Territory).

Mulgara, *Dasycercus cristicauda* (photo by H.J. Aslin).

Program) and NFCAP, a special educational program be developed on the principles of Wildlife Management Areas to promote better understanding amongst landowners of the vital role that such areas play in the conservation of wildlife.

Recommendation 6
This Action Plan recommends that no environmentally destructive development projects should be permitted to take place in any protected area.

Recommendation 7
This Action Plan recommends that the Papua New Guinea government should make a strong commitment to the urgent implementation, within the NFCAP framework, of this "Emergency Strategy" for the benefit of marsupial and monotreme conservation.

The following component urges the national government to overhaul the existing conservation machinery in Papua New Guinea, with a view to designing an effective and practicable conservation system. This component recognises a similar intent under the auspices of NFCAP, but stresses the need for a detailed consideration of the conservation requirements of marsupials and monotremes under such a review:

Recommendation 8
This Action Plan recommends that the Papua New Guinea government consider the establishment of a Nature Conservation Authority. This body would be the most senior environmental agency directing conservation programs through other existing conservation organisations and bodies.

Recommendation 9
This Action Plan recommends the comprehensive review of all conservation legislation, particularly the Conservation Areas Act (1978), with particular concern and reference to the substantial improvement of laws governing the management and protection of threatened species.

Recommendation 10
This Action Plan recommends that the Papua New Guinea government should enact threatened species legislation that protects both species and their critical habitats.

Recommendation 11
This Action Plan recommends that the Papua New Guinea Department of Environment and Conservation should pursue cooperative arrangements with conservation bodies, both international and national, to seek expertise and resources for the development of Wildlife Management Areas and other protected areas, and also the implementation of a comprehensive species conservation program.

Recommendation 12
This Action Plan recommends that early assessment for Wildlife Mamagement Areas should take account of the protected area proposals outlined by Diamond (undated) and Schodde (1973), and include research and integration of the conservation needs of marsupials and monotremes in the final determination of the national protected area system.

It is suggested that the determination of protected areas should be based on an assessment of the following: (a) high biological and ecological significance; (b) high human population densities; (c) the vicinity of major developmental activities. Scientific mechanisms to account for the needs of threatened marsupials and monotremes should be provided.

Recommendation 13
This Action Plan recommends that the Papua New Guinea government should make a firm commitment to provide sufficient staff, funds, and working facilities necessary for the development of a comprehensive species conservation program with initial emphasis on the management of marsupials and monotremes. Program components should include:

- the establishment of a Species Survival Commission National Committee under the chairmanship of the Secretary of the Department of Environment and Conservation to oversee all research and management programs for threatened species, and to ensure enforcement of endangered species legislation. The committee should be established in cooperation with other line departments. The Species Survival Commission would be seen as a technical advisory committee to the proposed Nature Conservation Authority.

- the government should make a strong commitment to strengthen, develop, and continue the species database inventory and atlas work for marsupials and monotremes, currently underway within the Department of Environment and Conservation. This should be integrated into any future GIS systems established under the NFCAP.

- The Department of Environment and Conservation, through the SSC National Committee, should, in cooperation with national and international experts, seek to confirm the status of all threatened marsupials and monotremes. This should also be seen as an important component of any national biological diversity survey conducted under the NFCAP.

Banded hare-wallaby, *Lagostrophus fasciatus* (photo by J. Short/CSIRO-Wildlife and Ecology).

- The SSC National Committee, Department of Environment and Conservation, NGOs, the Education Department, and other line departments should establish comprehensive publicity and education programs centred on the conservation of threatened species. Educational components should also be directed at government departments and non-government organisations.

Recommendation 14
This Action Plan recommends that the conservation actions identified for Endangered and Vulnerable species be vigorously pursued by the National Government.

Recommendation 15
This Action Plan recommends that research into solving the problems associated with determining precise taxonomic relationships for Papua New Guinea mammals generally, and marsupials in particular, should not be given any precedence over immediate and critical conservation requirements.

Recommendation 16
This Action Plan recommends that strenuous efforts be made to improve communications between government agencies, including access to computer databases (in Australia and Papua New Guinea), providing increased information exchange within the Department of Environment and Conservation.

Recommendation 17
This Action Plan recommends that the Papua New Guinea government's National Forest and Conservation Action Plan (NFCAP) allocate specific resources for significant increases in wildlife research and management efforts generally, and for the implementation of this Action Plan specifically.

Recommendation 18
This Action Plan recommends that any Papua New Guinea marsupial or monotreme with a limited geographic or altitudinal range should be considered "Potentially Vulnerable". All Potentially Vulnerable species identified in this Action Plan should be monitored on as regular a basis as possible.

Recommendation 19
This Action Plan recommends that surveys to establish the status and conservation needs of the endemic marsupials on Papua New Guinea islands be undertaken as a first step towards a conservation program.

Specific Recommendations

Recommendation 20
This Action Plan recommends the reservation and protection of known populations of the long-beaked echidna (*Zaglossus bruijni*), combined with public education programs and the investigation of captive breeding program potential.

Recommendation 21
This Action Plan recommends surveys to establish the status and ecology of the black-spotted cuscus (*Spilocuscus rufoniger*), and the implementation of a conservation regime to ensure its long-term survival.

Recommendation 22
This Action Plan recommends the reservation and protection of significant populations of Doria's tree-kangaroo (*Dendrolagus dorianus*).

Recommendation 23
This Action Plan recommends surveys to establish the range, status, and ecology of Goodfellow's tree-kangaroo (*Dendrolagus goodfellowi*), the reservation and protection of significant populations of this species, and participation in the captive breeding program.

Recommendation 24
This Action Plan recommends surveys to establish the status and ecology of the Woodlark Island cuscus (*Phalanger lullulae*), and the reservation and protection of viable populations of this species.

Recommendation 25
This Action Plan recommends surveys in the Telefomin area of western Papua New Guinea to establish the range, status, and ecology of the great-tailed triok (*Dactylopsila megalura*), and the Telefomin cuscus (*Phalanger matanim*).

Recommendation 26
This Action Plan recommends that surveys be undertaken to establish the range, status, and ecology of Scott's tree-kangaroo (*Dendrolagus scottae*) and the northern glider (*Petaurus abidi*), species endemic to the northern coastal ranges of Papua New Guinea.

Recommendation 27
This Action Plan recommends surveys in the southeast hill forests of Papua New Guinea to determine the range, status, and ecology of the giant bandicoot (*Peroryctes braodbenti*) and Macleay's dorcopsis (*Dorcopsis macleayi*).

General Recommendations for Australia

Recommendation 1

Noting that

- the Australian Constitution provides that land-use and nature conservation responsibilities reside mainly in the states and territories, with the Commonwealth government controlling exports and imports and being able to exercise its external affairs powers to intervene in some land-use decisions; and
- special threatened species legislation is required at all levels of government if national goals are to be realised;

this Action Plan recommends that all Australian governments enact appropriate threatened species legislation that protects threatened species and their habitats. Such legislation should include provision for written management and recovery plans for Endangered and Vulnerable species with further provisions for public participation.

Recommendation 2

Noting that

- the prevention of further extinctions of Australian marsupials will require detailed prescriptions for the management and recovery of all Endangered and Vulnerable species; and
- many species occur in more than one state or territory;

this Action Plan recommends to all states and territories that written Recovery Plans be prepared for all species of marsupials listed in this Plan as Endangered or Vulnerable and that where a species occurs within more than one state or territory, Interstate Recovery Teams be set up to carry out and coordinate the necessary work.

Recommendation 3

Noting that

- research and management programs for many of the Endangered and Vulnerable marsupials identified in this Plan should be, in most cases, long-term in nature;
- reintroduction projects will in most cases have at least a three- to five-year time scale; and
- predator control, fire management, and other habitat management activities will be a continuing requirement for the conservation of most of these species;

this Action Plan recommends that federal, state, and territory governments make firm commitments to provide sufficient staff and funds for carrying out the necessary work.

Long-footed potoroo, *Potorous longipes* (photo by J.H. Seebeck, Victorian Department of Conservation and Environment).

Recommendation 4

Noting that

- some islands contain especially important populations of threatened marsupials (Tables17 and 18-Chapter 4);

this Action Plan recommends that all islands with populations of Endangered, Vulnerable, or Potentially Vulnerable marsupials be protected from damaging actions and that management plans be prepared for them.

Recommendation 5

Noting that

- species recovery plans are unlikely to succeed without community support; and
- endangering processes are likely to continue unless community attitudes change;

this Action Plan recommends that education programs that stress the effects of habitat loss and degradation on the conservation status of marsupials and monotremes be designed and implemented in both formal and informal education arenas.

Recommendation 6

Noting that

- the European red fox is identified in this Plan as one of the major causes of marsupial extinction and decline;
- some states are already carrying out research into fox biology and control, and are controlling fox numbers around remnant populations of endangered and vulnerable marsupials;
- the Commonwealth has committed funds to support a fox control research program; and
- foxes are a national problem and long-term eradication or effective control will only be possible if the problem is approached nationally;

this Action Plan recommends that a national fox control program be established and that this include research into biological control methods, more cost-effective local control methods, and the ecology of predation.

Recommendation 7

Noting that

- several Australian marsupial species are regarded as Vulnerable or Potentially Vulnerable because their rain forest habitat is limited in extent; and
- many areas of rainforest in Australia do not have the highest level of protection under both State and Federal legislation and could potentially be easily eliminated;

this Action Plan recommends to governments that rainforests of importance to the conservation of Vulnerable and Potentially Vulnerable marsupials be afforded the highest level of protection.

Recommendation 8

Noting

- the particular susceptibility of "critical weight range" (CWR) species; and
- the relatively small research effort being directed towards

these species compared to the large kangaroos and wallabies and the small dasyurids;

this Action Plan recommends that nature conservation and research fund-granting agencies redirect their research effort and grants to CWR species.

Recommendation 9

Noting that

- some research activities have general application to a large number of species and action in these fields will provide a useful catalytic effect and general benefit to conservation of endangered species;
- this plan has identified causes of marsupial decline and extinction common to many species, viz.
  - habitat destruction (i.e. deforestation, conversion to other non-natural ecosystems, urbanization, etc.) and habitat fragmentation;
  - habitat degradation/modification, commonly caused by exotic herbivores (rabbits, goats, etc.) and inappropriate fire regimes; and
  - predation by introduced species, especially the European red fox and feral cat;
- the effect of two or more of these causes in combination may be much more severe than would be expected by simple addition;
- combating these processes/causes is essential to prevent further marsupial extinctions and to reverse declines that have already occurred; and
- overcoming these processes will lead to the effective conservation of many species;

this Action Plan recommends that funds be allocated to identifying and combating processes or groups of processes as well as for the conservation of individual species. It should be noted that projects that address a combination of processes or species are likely to be more cost-effective than those that address a single process or species.

Recommendation 10

Noting that

- fragmentation of habitat results in the genetic isolation of subpopulations of those threatened species which are unable to disperse;
- the risk of such isolated populations losing genetic diversity due to random loss of alleles (genetic drift) is high;
- such loss of allelic diversity reduces each isolated population's chances of adapting to change or surviving stochastic environmental events;
- every loss of an isolated population depletes the genetic diversity of the species as a whole;

this Action Plan recommends that management agencies make provision, in species recovery and management plans for threatened species, for the preservation of existing genetic diversity and the maintenance of long-term evolutionary potential, by:

1. maximixing the size of isolated populations to minimise the risk of short-term genetic drift or loss through catastrophic stochastic events;
2. managing migration between populations to replace lost opportunities for gene flow; and
3. relocation, in suitable alternative habitat or in captivity, of populations which have a low probability of survival in the short term.

Recommendation 11

Noting that

- to provide a better definition of management requirements and develop cost-effective management techniques additional research is needed into:
  - decline and recovery of marsupial populations during and after periods of stress, e.g. drought;
  - the development of cost-effective survey and monitoring techniques for marsupial populations; and
  - the development of techniques for translocation and reintroduction of populations;

this Action Plan recommends that nature conservation and research fund-granting agencies give priority to such studies.

Recommendation 12

Noting that it has been suggested that the survival of many arid zone mammals is dependent on habitats of "exceptionally high quality in terms of the dependability and nutritional character of their plant growth" (Morton 1990), this Action Plan recommends that a project to test this hypothesis be designed and funded, possibly involving the common brushtail possum (*Trichosurus vulpecula*) in central Australia.

Brush-tailed rock-wallaby, *Petrogale persephone* (photo by J.R.D. Bartlett).

Specific recommendations for all Endangered and Vulnerable Australian marsupial species are given in Appendix II.

Recommendations that are not Country-Specific

Recommendation 1

This Action Plan recommends that this Action Plan be reviewed by the IUCN Species Survival Commission Australasian Marsupial and Monotreme Specialist Group (AMMSG) after five years, especially regarding the implementation of its recommendations and any significant changes in status of monotremes and marsupials.

Recommendation 2

This Action Plan recommends that Action Plans similar to this one be developed for other vertebrate groups in Australasia, commencing with Action Plans for native rodents and bats in both Australia and New Guinea.

Golden bandicoot, *Isoodon auratus* (photo by Andrew Burbidge, Department of Conservation and Land Management, Western Australia).

Appendix 1: A List of the Monotremes and Marsupials of the New Guinea Area

Compiled by Graeme G. George for the

IUCN/SSC Australasian Marsupial and Monotreme Specialist Group

This list covers the marsupials of Indonesia, Papua New Guinea and the Solomon Islands, and the monotremes of New Guinea, showing the latest taxonomy accepted by those working with this fauna. When this list was commenced the handbooks of Flannery (1990a) and Menzies (1991), covering the New Guinea fauna, had not been published. This list extends the coverage and updates these recent publications, both of which were delayed in press.

The last full checklist of the mammals of the area is that of Laurie & Hill (1954). This list brings together all the changes to the taxonomy of monotreme and marsupial species published since 1954.

Primary references are shown bold in the list. References for newly-described species show only the original author/s, where the new species has been accepted by subsequent workers. It is clear that much taxonomic work remains to be done on the marsupials of this area. This list attempts to indicate the areas of instability and divergence of opinion which may result in future changes.

Monotremata

Family Tachyglossidae

Tachyglossus aculeatus, Short-beaked echidna, (Shaw & Nodder 1792)

Zaglossus bruijni, Long-beaked echidna, (Peters & Doria 1876) [includes *bartoni* and *bubuensis*]—**van Deusen & George (1969)**, Griffiths (1978), Ziegler (1982), Honacki et. al (1982), Corbett & Hill (1986), Flannery (1990a), Menzies (1991).

Marsupialia

Family Dasyuridae

Neophascogale lorentzi, Speckled dasyure, (Jentink 1911)

Murexia longicaudata, Short-furred dasyure, (Schlegel 1866)

"Murexia" rothschildi, Broad-striped dasyure, Tate 1938 [congeneric with *melanurus*—van Dyck (1988b)]

"Antechinus" melanurus (Thomas 1899), Black-tailed antechinus [includes *wilhelmina*]—van Dyck (1988b), however Flannery (1990a) and Menzies (1991) maintain *wilhelmina* as a separate species.

"Antechinus" habbema, Lesser antechinus, Tate & Archbold 1941 [generically distinct from other small dasyurids—van Dyck (1988b). See also Woolley (1984, 1989).]

"Antechinus" naso, Long-nosed antechinus, (Jentink 1911) [generically distinct from other small dasyurids—van Dyck (1988b). See also Woolley (1984, 1989).]

Phascolosorex doriae, Red-bellied dasyure, (Thomas 1896)

Phascolosorex dorsalis, Narrow-striped dasyure, (Peters & Doria 1876)

Myoictis melas, Three-striped dasyure, (Muller 1840)

Planigale novaeguineae, Papuan planigale, Tate & Archbold 1941

Dasyurus albopunctatus, New Guinea quoll, Schlegel 1880 —Ride (1970) placed *Satanellus hallucatus* in *Dasyurus* and most subsequent authors have applied *Dasyurus* in this wider sense.

Dasyurus spartacus, Bronze quoll, van Dyck 1988 —**van Dyck (1988a).**

Sminthopsis archeri, Chestnut dunnart, van Dyck 1986 —**van Dyck (1986).**

Sminthopsis virginiae, Red-cheeked dunnart, Tarragon 1847 [includes *rufigenis*]—**Archer (1981)** and subsequent authors.

Family Peroryctidae—Groves & Flannery (1990)

Peroryctes broadbenti, Broadbent or Giant bandicoot, (Ramsay 1879) —van Deusen & Jones (1967), van Deusen (1972), Honacki et. al. (1982), Corbett & Hill (1986), George & Maynes (1990), **Groves & Flannery (1990)**, Flannery (1990a), Menzies (1991). [Not recognized by Kirsch & Calaby (1977). Ziegler (1982) gives qualified recognition.]

Peroryctes raffrayana, Raffray bandicoot, (Milne-Edwards 1878) [Spelling of *raffrayanus* emendated by Flannery (1990a) to match the gender of *Peroryctes* under Article 31 of the International Code.]

Microperoryctes longicauda, Striped bandicoot, (Peters & Doria 1876) —**Groves & Flannery (1990a)**, Flannery (1990a).

Microperoryctes murina, Mouse bandicoot, Stein 1932

Microperoryctes papuensis, Papuan bandicoot, (Laurie 1952) —**Groves & Flannery (1990a)**, Flannery (1990a).

Echymipera clara, Clara or Dimorphic bandicoot, Stein 1932

Echymipera echinista, Fly River bandicoot, Menzies 1990 —**Menzies 1990.**

Echymipera davidi, Kiriwina bandicoot, Flannery 1990 —Flannery (1990a), **Flannery (1990b).**

Echymipera kalubu, Kalubu or Black-headed bandicoot, (Fischer 1929) [Fischer is shown as the author of *kalubu* following Husson (1955)].

Echymipera rufescens, Rufous bandicoot, (Peters & Doria 1875)

Rhynchomeles prattorum, Seram bandicoot, Thomas 1920

Family Peramelidae

Isoodon macrourus, Northern brown or Brindled bandicoot, (Gould 1842) —Marlow (1960), **Lidicker & Follett (1968)**, Ride (1970) and all subsequent authors.

Family Macropodidae

Dorcopsis atrata, Black dorcopsis/wallaby, Van Deusen 1955 —**Van Deusen (1955).**

Dorcopsis hageni, White-striped dorcopsis wallaby, Heller 1897

Dorcopsis luctuosa, Grey dorcopsis wallaby, (D'Albertis 1874) —Archer, Flannery & Grigg (1985), **Groves & Flannery (1989)**, Menzies (1989), Flannery (1990a), Menzies (1991).

Dorcopsis muelleri, Brown dorcopsis wallaby, (Schlegel 1866) —George & Schurer (1978), Corbett & Hill (1986), **Groves & Flannery (1989)**, Menzies (1991). [Other authors have continued to use *veterum*. Flannery (1990a) uses *veterum* but this publication was delayed in press and Groves & Flannery (1989) is a later opinion.]

Dorcopsulus macleayi, Macleay's dorcopsis wallaby, (Mikluho-Maclay 1885)

Dorcopsulus vanheurni, Little dorcopsis wallaby, (Thomas 1892)

Dendrolagus dorianus, Doria's tree-kangaroo, Ramsay 1883

Dendrolagus goodfellowi, Goodfellow's tree-kangaroo, Thomas 1908 [Groves (1982) placed this taxon in *matschiei*, as have some previous authors. Menzies (1991) accepts Groves's arrangement, but Procter-Gray and Ganzlosser (1986), Flannery (1990a), and most other authors have continued to recognize *goodfellowi* as a full species.]

Dendrolagus inustus, Grizzled tree-kangaroo, Muller 1840

Dendrolagus matschiei, Matschie's tree-kangaroo, Forster & Rothschild 1907 [includes *deltae*—Lidicker & Ziegler (1968), Groves (1982)].

Dendrolagus scottae, Scott's tree-kangaroo, Flannery & Seri 1990 —**Flannery & Seri (1990)**

Dendrolagus spadix, Lowland tree-kangaroo, Troughton & Le Soeuf 1936 [synonymized with *goodfellowi* by most authors since Lidicker & Ziegler (1968) but recognized as a full species by Maynes (1989) and Flannery (1990a). Groves (1982) and Menzies (1991) include this taxon in *matschiei*.]

Dendrolagus ursinus, Black or White-throated tree-kangaroo, Muller 1840

Thylogale brunii, Brown or Dusky pademelon, (Schreber 1778) [the original spelling (*brunii* not *bruijni*) has been used by most authors subsequent to Husson (1955).]

Thylogale stigmatica, Red-legged pademelon, (Gould 1860)

Macropus agilis, Agile callaby, (Gould 1842) —**Stirton (1963)**, Ride (1970) and all subsequent authors.

Family Phalangeridae

Ailurops ursinus, Large Celebes cuscus, (Temminck 1824) —**Flannery et al. (1987), George (1987).**

Strigocuscus celebensis, Little Celebes cuscus, (Gray 1858) —**Flannery et al. (1987), George (1987).**

Phalanger pelengensis, Peleng Island cuscus, Tate 1945 —**Flannery et al. (1987), George (1987), Groves (1987b).** [Flannery et. al (1987) suggest this species may be generically distinct from other phalangerids.]

Phalanger gymnotis, Ground cuscus, (Peters & Doria 1875) [Van der Feen (1962), Feiler (1978a), Honacki et. al (1982) and George (1987) distinguished *leucippus* of New Guinea from *gymnotis* of the Aru Islands. Flannery (1990a), Flannery & Seri (1990) and Menzies (1991) include *leucippus* in *gymnotis*. For the purposes of the Action Plan the two forms are distinguished at the subspecific level as their conservation status may be quite different.]

Phalanger ornatus, Moluccan cuscus, Gray 1860 —George (1982), **Menzies & Pernetta (1986), George (1987), Groves (1987b).** [May be a complex of more than one species—Flannery (unpubl.), George (unpubl.)]

Phalanger rothschildi, Obi Island cuscus, Thomas 1898 **—Menzies & Pernetta (1986), George (1987), Groves (1987b).** [Flannery et. al (1987) place *ornatus*, (*rothschildi*) and *gymnotis* in Strigocuscus, but Flannery (pers. comm.) no longer supports this arrangement.]

Phalanger carmelitae, Mountain cuscus, Thomas 1898 —George (1979), George (1982), Ziegler (1982), Honacki et. al (1982), Corbett & Hill (1986), **Menzies & Pernetta (1986), Flannery et al. (1987), George (1987), Groves (1987a)**, Flannery (1990a), Menzies (1991).

Phalanger lullulae, Woodlark Island cuscus, Thomas 1896 —George (1979), George (1982), Honacki et. al (1982), Ziegler (1982), Corbett & Hill (1986), **Menzies & Pernetta (1986), Flannery et al. (1987), George (1987)**, Menzies (1991).

Phalanger matanim, Telefomin cuscus, Flannery 1987 **—Flannery (1987).**

Phalanger orientalis, Grey or Common cuscus, (Pallas 1766) [includes *mimicus* sensu Flannery et al. (1987), however electrophoretic data confirms that *P. orientalis* is a complex of at least three species—Flannery (pers. comm.). See also George (1987).]

Phalanger sericeus, Silky cuscus, Thomas 1907 —**Menzies & Pernetta** (1986), Flannery (1990a), Menzies (1991). [George (1979/82/87 included this taxon in *vestitus*].

Phalanger vestitus, Stein's cuscus, Milne-Edwards 1877 [includes *P. interpositus* (sensu George 1979/1982/1987) & *P. permixtio* Menzies & Pernetta, 1986]—**Menzies & Pernetta (1986)**, Flannery (1990a), Flannery & Seri (1990). Menzies (1991) maintains *permixtio* as a full species.]

Spilocuscus kraemeri, Manus Island spotted puscus, (Schwarz 1910) —Flannery et al. (1984), Flannery (unpubl.), George (unpubl.). See also Flannery & Calaby (1987).

Spilocuscus maculatus, Common spotted puscus, (Geoffroy 1803) —**Flannery et al. (1987)**, Flannery & Calaby (1987), **George (1987)**, Flannery (1990a), Menzies (1991).

Spilocuscus papuensis, Waigeo Island spotted cuscus, (Desmarest 1822) —**George (1987)**, Menzies (1991)

Spilocuscus rufoniger, Black-spotted cuscus, (Zimara 1937) [includes *atrimaculatus*]—Feiler (1978b), George (1979), George (1982), Ziegler (1982), Honacki et al. (1982), Corbett & Hill (1986), **Flannery et al. (1987), George (1987)**, Flannery (1990a), Menzies (1991).

Family Acrobatidae—Aplin (1987)

Distoechurus pennatus, Feather-tailed possum, (Peters 1874)

[*Acrobates pulchellus*, New Guinea feathertail glider, Rothschild 1893]

Family Burramyidae—Kirsch (1968)

Cercartetus caudatus, Long-tailed pygmy-possum, (Milne-Edwards 1877) —**Wakefield (1963)** and all subsequent authors.

Family Petauridae—Kirsch (1968), Archer (1984)

Petaurus abidi, Torricelli or Northern glider, Ziegler 1981 —**Ziegler (1981)**.

Petaurus breviceps, Sugar glider, Waterhouse 1838

Dactylopsila megalura, Great-tailed triok, Rothschild & Dollma 1932

Dactylopsila palpator, Long-fingered triok, Milne-Edwards 1888 —Kirsch & Calaby (1977), Ziegler (1982), Honacki et. al (1982), Corbett & Hill (1986), Flannery (1990a), Menzies (1991).

Dactylopsila tatei, Fergusson Island striped possum, Laurie 1952

Dactylopsila trivirgata, Common striped possum, Gray 1858

Family Pseudocheiridae—Archer (1984), Smith (1984)

Pseudochirops cupreus, Copper ringtail, (Thomas 1897) —Archer (1984), Flannery (1990a).

Pseudochirops albertisi, D'Albertis ringtail, (Peters 1874) —Archer (1984), Flannery (1990a).

Pseudochirops corinnae, Golden or Plush-coated ringtail, (Thomas 1897) —Archer (1984), Flannery (1990a).

Pseudocheirus caroli, Weyland ringtail, Thomas 1921

Pseudocheirus forbesi, Painted ringtail, Thomas 1897

Pseudocheirus schlegeli, Arfak ringtail, Jentink 1884

Pseudocheirus mayeri, Pygmy ringtail, Rothschild & Dollman 1932

Pseudocheirus canescens, Lowland or Hoary ringtail, (Waterhouse 1846)

References to Appendix 1

Aplin, K.P. 1987. Basicranial anatomy of the early Miocene diprotodontan *Wynyardia bassiana* (Marsupialia: Wynyardiidae and its implications for wynyardiid phylogeny and classification. In: *Possums and Opossums: Studies in Evolution,* ed. by M. Archer. Royal Zoological Society of New South Wales: Sydney. Pp. 369-91.

Archer, M. 1981. Systematic revision of the marsupial genus *Sminthopsis* Thomas. *Bull. Amer. Mus. Nat. Hist.* 168:61-224.

—. 1984. The Australian marsupial radiation. In: *Vertebrate zoogeography and evolution in Australasia,* ed. by M. Archer and G. Clayton. Hesperian Press: Perth. Pp. 633-809.

Archer, M., Flannery, T.F. and Grigg, G.C. 1985. *Kangaroos.* Kevin Weldon: Sydney.

Corbett, G.B., and Hill, J.E. 1986. *A world list of mammalian species, Second Edition.* British Museum (Natural History): London.

Feiler, A. 1978a. Bemerkungen uber *Phalanger* der *"orientalis-*Gruppe" nach Tate (1945) (Mammalia, Marsupialia). *Zool. Abhandlungen Mus. Tierk. Dresden.* 34: 385-95.

—. 1978b. Uber artliche abgrenzung und innerartliche ausformung bei *Phalanger maculatus* (Mammalia, Marsupialia, Phalangeridae. *Zool. Abhandlungen Mus. Tierk. Dresden.* 35:1-30.

•Flannery, T. 1987. A new species of *Phalanger* (Phalangeridae: Marsupialia) from montane western Papua New Guinea. *Rec. Aust. Mus.* 39:183-193.

Flannery, T.F. 1990a. *The Australian Museum Mammals of New Guinea.* Robert Brown and Associates: Queensland.

—. 1990b. *Echymipera davidi,* a new species of Perameliformes (Marsupialia) from Kiriwina Island, Papua New Guinea, with notes on the systematics of the genus *Echymipera.* In: *Bandicoots and Bilbies,* ed. by J. S. Seebeck, P. R. Brown, R. Wallis and C. Kemper. The Australian Mammal Society/Surrey Beatty and Sons: Sydney.

Flannery, T.F., Archer, M., Baverstock, P., Maynes, G. and Aplin, K. 1984. Interrelationships of the Phalangeridae. (Abstract) *Bull. Aust. Mamm. Soc.* 8:109.

Flannery, T., Archer, M., and Maynes, G. 1987. The phylogenetic relationships of living Phalangerids (Phalangeroidea: Marsupialia) with a suggested new taxonomy. In: *Possums and Opossums: Studies in Evolution,* ed. by M. Archer. Royal Zoological Society of New South Wales: Sydney. Pp. 477-506.

Flannery, T., and Calaby, J.H. 1987. Notes on the species of *Spilocuscus* (Marsupialia: Phalangeridae) from northern New Guinea and the Admiralty and St. Matthias Island groups. In: *Possums and Opossums: Studies in Evolution,* ed. by M. Archer. Royal Zoological Society of N.S.W. Sydney. Pp. 547-58.

•Flannery, T.F. and Seri, L. 1990. *Dendrolagus scottae* n. sp. (Marsupialia: Macropodidae): a new Tree-kangaroo from Papua New Guinea. *Rec. Aust. Mus.* 42: 237-45.

•—. 1990b. The mammals of southern West Sepik Province, Papua New Guinea: their distribution, abundance, human use and zoogeography. *Rec. Aust. Mus.* 42: 173-208.

George, G.G. 1973. Land mammal fauna. *Aust. Nat. Hist.* December 1973: 420-426. Reprinted with minor revisions as "The present day land mammal fauna of Papua New Guinea". In: *Vertebrate Zoogeography and Evolution in Australasia,* ed. by M. Archer and G. Clayton. Hesperian Press: Perth, 1984. Pp. 1137-1141.

—. 1979. The status of endangered Papua New Guinea mammals. In: *The status of endangered Australasian wildlife,* ed. by M. Tyler. Royal Zoological Society of South Australia: Adelaide. Pp. 93-100.

—. 1982. Cuscuses *Phalanger* spp.: their management in captivity. In: *The management of Australian mammals in captivity,* ed. by D.D. Evans. Zoological Board of Victoria: Melbourne. Pp. 67-72.

—. 1987. Characterisation of the living species of Cuscus (Marsupialia: Phalangeridae). In: *Possums and Opossums: Studies in Evolution,* ed. by M. Archer. Royal Zoological Society of N.S.W.: Sydney. Pp. 507-526.

George, G.G., and Schurer, U. 1978. Notes on macropods commonly mis-identified in zoos. *Int. Zoo. Yearb.* 18:152-156.

George, G.G., and Maynes, G.M. 1990. The Status of New Guinea Bandicoots. In: *Bandicoots and Bilbies,* ed. by J.H. Seebeck, P.R. Brown, R. Wallis and C. Kemper. The Australian Mammal Society/ Surrey Beatty and Sons: Sydney. Pp. 93-105.

Griffiths, M. 1978. *The Biology of the Monotremes.* Academic Press: New York. Pp. 71.

Groves, C.P. 1982. The systematics of tree-kangaroos (*Dendrolagus;* Marsupialia, Macropodidae). *Aust. Mammal.* 5:156-186.

Groves, C. 1987a. On the highland cuscuses (Marsupialia: Phalangeridae) of New Guinea. In: *Possums and Opossums: Studies in Evolution,* ed. by M. Archer. Royal Zoological Society of N.S.W.: Sydney. Pp. 559-67.

—. 1987b. On the cuscuses (Marsupialia: Phalangeridae) of the *P. orientalis* group from Indonesian Territory. In: *Possums and Opossums: Studies in Evolution,* ed. by M. Archer. Royal Zoological Society of N.S.W: Sydney. Pp. 569-79.

Groves, C.P., and Flannery, T.F. 1989. Revision of the genus *Dorcopsis* (Macropodididae: Marsupialia). In: *Kangaroos, wallabies and rat-kangaroos,* ed. by G. Grigg, P. Jarman and I. Hume. Surrey Beatty and Sons: Sydney. Pp. 117-128.

—. 1990. Higher level systematics within the Peramelidae (Marsupialia). In: *Bandicoots and Bilbies,* ed. by J.S. Seebeck, P.R. Brown, R. Wallis and C. Kemper. The Australian Mammal Society/Surrey Beatty and Sons: Sydney.

Husson, A.M. 1955. Notes on the mammals collected by the Swedish New Guinea Expedition 1948-49. *Nova Guinea* n. s. 6:283-306.

Kirsch, J.A.W. 1968. Prodromus of the comparative serology of Marsupialia. *Nature* 217:418-420.

Kirsch, J.A.W., and Calaby, J.H. 1977. The species of living marsupials: an annotated list. In: *The Biology of Marsupials,* ed. by B. Stonehouse and D. Gilmore. Macmillan Press: London. Pp. 9-26.

Laurie, E.M.O., and Hill, J.E. 1954. *List of land mammals of New Guinea, Celebes and Adjacent Islands 1758-1952.* British Museum (Natural History): London.

Lidicker, W.Z., and Follett, W.I. 1968. *Isoodon* Desmarest, 1817, rather than *Thylacis* Illiger, 1811, as the valid generic name of the short-nosed bandicoots (Marsupialia: Peramelidae). *Proc. Biol. Soc. Washington* 81:251-256.

Lidicker, W.Z., Jr., and Ziegler, A.C. 1968. Report on a collection of mammals from eastern New Guinea. *Univ. Calif. Publ. Zool.* 87:1-64.

Marlow, B.J. 1960. The generic name of the short-nosed bandicoots. *Aust. Mamm. Soc. Bull.* 2:12.

Maynes, G.M. 1989. Zoogeography of the Macropodoidea. In: *Kangaroos, Wallabies and Rat-kangaroos,* ed. by G. Grigg, P. Jarman and I. Hume. Surrey Beatty and Sons: Sydney. Pp. 47-66.

•Menzies, J.I. 1990. Notes on Spiny Bandicoots, *Echymipera* spp. (Marsupialia: Peramelidae) from New Guinea and description of a new species. *Science in New Guinea* 16 (2) 86-98.

•Menzies, J.I., and Pernetta, J.C. 1986. A taxonomic revision of the cuscuses allied to *Phalanger orientalis* (Marsupialia: Phalangeridae). *J. Zool., Lond.* (B) 1:551-618.

Proctor-Gray, E., and Ganzlosser, U. 1986. The individual behaviours of Lumholtz's Tree-kangaroo: repertoire and taxonomic implications. *J. Mammal.* 67:343-352.

Ride, W.D.L. 1970. *A Guide to the Native Mammals of Australia.* Oxford University Press: Melbourne.

Smith, A. 1982. The species of living possums and gliders. In: *Possums and Gliders,* ed. by A.P. Smith and I. Hume. Australian Mammal Society/Surrey Beatty and Sons: Sydney. Pp. xiii-xv.

Stirton, R.A. 1963. A review of the macropodid genus *Protemnodon. Univ. Calif. Publ. Geol.* 44:97-162.

van der Feen, P.J. 1962. Catalogue of the Marsupialia from New Guinea, the Moluccas and Celebes of the Museo Civico di Storia Naturale "Giacomo Doria" in Genoa. *Ann. Mus. Civ. Storia Natur. Genova* 73:19-70.

van Deusen, H.M. 1972. Marsupials. In: *Encyclopaedia of Papua and New Guinea,* ed. by P. Ryan. Melbourne University Press/University of Papua and New Guinea: Melbourne. Pp. 711-714.

—. 1955. Results of the Archbold Expeditions. No. 76. A new species of wallaby (genus *Dorcopsis*) from Goodenough Island, Papua. *Am. Mus. Novitates* 1826:1-25.

van Deusen, H.M. and George, G.G. 1969. Results of the Archbold Expeditions No. 90. Notes on the echidnas (Mammalia, Tachyglossidae) of New Guinea. *Am. Mus. Novitates* 2383.

van Deusen, H.M. and Jones, J.K., Jr. 1967. Marsupials. In: *Recent Mammals of the World: a synopsis of families,* ed. by S. Anderson, and J. Knox Jones, Jr. Ronald Press: New York.

van Dyck, S.M. 1986. The Chestnut Dunnart, *Sminthopsis archeri* (Marsupialia: Dasyuridae), a new species from the savannas of Papua New Guinea and Cape York Peninsula, Australia. *Aust. Mammal* 9:111-124.

—. 1988a. The Bronze Quoll, *Dasyurus spartacus* (Marsupialia: Dasyuridae), a new species from the savannas of Papua New Guinea. *Aust. Mammal* 11:145-156.

—. 1988b. Phylogenetics of Papua New Guinean and Irian Jayan Dasyrurids (Abstract). Programme and Abstracts, 1988 AGM, Australian Mammal Society.

Wakefield, N.A. 1963. The Australian Pygmy-possums. *Victorian Nat.* 80:99-116.

Woolley, P.A. 1984. Phallic morphology of the New Guinean species of *Antechinus.* (Abstract). *Bull. Aust. Mamm. Soc.* 8:182.

—. 1989. Nest location by spool and line tracking of dasyurid marsupials in New Guinea. *J. Zool., Lond.* 218: 689-700. Pp. 699.

Ziegler, A.C. 1981. *Petaurus abidi,* a new species of glider (Marsupialia: Petauridae) from Papua New Guinea. *Aust. Mammal* 4:81-88.

—. 1982. An ecological checklist of the New Guinea recent mammals. In: *Biogeography and ecology of New Guinea,* ed. by J.L. Gressitt. W. Junk: The Hague. Pp. 867-93.

Acknowledgements

This list was initiated with a request from Lester Seri for consensus on Papua New Guinea marsupial nomenclature. The assistance of Tim Flannery, Steve van Dyck, Pat Woolley, Jim Menzies, and Joan Dixon in providing opinions and copies of references is gratefully acknowledged.

Table 25. Index and costs summary.

	Species	Page No.	Costs $AU
1.	Numbat	68	800,000
2.	Kowari	69	640,000
3.	Mulgara	70	850,000
4.	Chuditch	71	325,000
5.	Dibbler	72	220,000+
6.	Red-tailed phascogale	73	270,000
7.	Julia Creek dunnart	74	50,000
8.	Sandhill dunnart	74	350,000
9.	Golden bandicoot	75	1,170,000
10.	Western barred bandicoot	76	1,540,000
11.	Eastern barred bandicoot	77	1,790,000
12.	Bilby	78	2,597,000
13.	Northern hairy-nosed wombat	79	642,000
14.	Boodie	80	1,793,000
15.	Brush-tailed bettong	81	470,000
16.	Northern bettong	83	480,000
17.	Long-footed potoroo	83	860,000
18.	Mala	85	1,830,000
19.	Banded hare-wallaby	86	1,160,000
20.	Tammar wallaby	87	560,000
21.	Bridled nailtail wallaby	88	294,000
22.	Black-footed rock-wallaby	89	920,000+
23.	Brush-tailed rock-wallaby	91	954,000
24.	Proserpine rock-wallaby	93	1,763,000
25.	Mountain pygmy-possum	93	1,145,000
26.	Western ringtail	95	200,000
27.	Leadbeater's possum	95	250,000
	Total		**A$23,923,000**

Species Recovery Outline No. 1

Family: Myrmecobiidae

Scientific name: *Myrmecobius fasciatus*

Common names: Numbat, Walpurti

Intra-specific taxa: None

Species survival status: Endangered

Former distribution: Widespread across southern semi-arid and arid Australia, from western New South Wales through South Australia and southern Northern Territory to the southwest of Western Australia. Did not occur on islands.

Current distribution: Very few, remnant populations in the southwest of Western Australia.

Habitat: Formerly in semi-arid and arid woodlands (*Eucalyptus* and *Acacia*) and grasslands (*Triodia* and *Plectrachne*). Now restricted to eucalyptus woodlands at wettest periphery of former range.

Reasons for decline: Predation by foxes (proven), changed fire regimes, especially in arid grasslands (speculative), habitat destruction in some areas (proven).

Conservation reserves on which species occurs: Perup Nature Reserve, Boyagin Nature Reserve, Karroun Hill Nature Reserve (all in Western Australia).

Other public lands on which species occurs: Dryandra State Forest, parts of State Forest in Northern, Central, and Southern Forest Regions of the Western Australian Department of Conservation and Land Management.

Other land on which species occurs: Areas of private and Commonwealth land near Jandakot Airport in Perth Metropolitan Area.

Is knowledge about species adequate for objectives and actions to be defined accurately? (If not, provide list of additional studies required): Yes.

Recovery Plan objectives:
- Retain current distribution
- Increase range by re-establishing in suitable areas.
- Maintain captive breeding colonies.

Management actions completed or underway:
- Most current habitat in conservation reserves; almost all current habitat in public lands.
- Fox control by poison baiting being carried out in Dryandra State Forest, Boyagin Nature Reserve, Karroun Hill Nature Reserve, Tutanning Nature Reserve.
- Translocation successfully completed to Boyagin Nature Reserve; currently underway to Karroun Hill Nature Reserve.

Management actions required:
- Declaration of additional conservation reserves: yes, parts of Dryandra State Forest should be converted to Nature Reserve or National Park.
- Habitat management: fire management required to conserve cover and log refuges. Modification of logging practices in numbat habitat in State Forest required where relevant.
- Feral animal control: control of foxes and feral cats essential. National program for fox control needed.
- Translocation: re-establish in Tutanning Nature Reserve and possibly other conservation reserves.
- Captive breeding: refinement of husbandry techniques required. Maintain existing colony at Perth Zoo and commence new captive colonies.
- Other: establishment of woodland corridors between conservation reserves may be required to provide areas of habitat large enough for viable populations. Existing (including translocated) populations require monitoring.

Organisations responsible for conservation of species: Western Australian Department of Conservation and Land Management (CALM).

Other organisation(s) or individual(s) involved: Perth Zoo. World Wide Fund for Nature Australia (WWF). Australian Geographic Society.

Conservation reserves affected by recovery plan (list other action plan species in each conservation reserve): Perup Nature Reserve:

Dasyurus geoffroii (chuditch), *Bettongia penicillata* (brush-tailed bettong), *Macropus eugenii* (Tammar wallaby), *Pseudocheirus occidentalis* (Western ringtail). Boyagin Nature Reserve: *Phascogale calura* (red-tailed phascogale), *Macropus eugenii* (Tammar wallaby). Karroun Hill Nature Reserve, Tutanning Nature Reserve: *Phascogale calura* (red-tailed phascogale), *Bettongia penicillata* (brush-tailed bettong), *Macropus eugenii* (Tammar wallaby).

Other public lands affected by conservation plan: Dryandra State Forest: *Phascogale calura* (red-tailed phascogale), *Bettongia penicillata* (brush-tailed bettong), *Macropus eugenii* (Tammar wallaby). State Forest in Conservation and Land Management (CALM) Northern, Central and Southern Forest Regions: *Dasyurus geoffroii* (chuditch), *Phascogale calura* (red-tailed phascogale).

Other lands affected by conservation plan: Areas of private and Commonwealth land near Jandakot Airport in Perth Metropolitan Area.

Can recovery plan be carried out with existing resources? (If not, list additional staff and financial resources required): No.

1. Fox control is required in large areas including Karroun Hill Nature Reserve and parts of jarrah forest until national program of control becomes effective.
   Cost: 1 staff ($A30,000) + $A50,000 costs per annum (x10)
   Total=$A800,000

   Note: Except fox and feral cat control research. National program to control fox commenced. Similar research on feral cat required.

Remarks: Extensive research carried out by the Western Australian Department of Conservation and Land Management with the financial assistance of World Wide Fund for Nature Australia and the Australian Geographic Society. Species management plan in preparation. Some parts of Dryandra State Forest that are inhabited by numbats are managed as though they are nature reserves.

Species Recovery Outline No. 2

Family: Dasyuridae

Scientific name: *Dasycercus byrnei*

Common name: Kowari

Intra-specific taxa: None

Species survival status: Endangered

Former distribution: As below but extending across the Northern Territory border and the northeastern quarter of South Australia.

Current distribution: South Australia/Northern Territory border, northern South Australia south and east of the Simpson Desert; Western Queensland between Simpson Desert and the Diamantina River south of Boulia.

Habitat: Gibber plains and gibber crests in undulating Mitchell grass plains.

Reasons for decline: Habitat destruction through grazing (speculative) fox predation (speculative).

Conservation reserves on which species occurs: Innamincka Regional Reserve, South Australia.

Other public lands on which species occurs: Grazing leases in northern South Australia, southeast Northern Territory and southwest Queensland.

Other land on which species occurs: None known.

Is knowledge about species adequate for objectives and actions to be defined accurately? (If not, provide list of additional studies required): No.

- Reserve and habitat management research on the species' ecological requirements (High priority).

Recovery Plan objectives:

- Survey current distribution.
- Reserve and or manage critical habitat.
- Research special/basic needs of species.

Management actions completed or underway:

- Research program distribution and habitat surveys (L. Lim).

Management actions required:

- Declaration of additional conservation reserves: yes, acquisition of habitat north of Birdsville between Diamantina Lakes and Coorbulka Stations. Management and protection of key colonies.
- Habitat management: retention of remnant habitats on private and public land. Establishment of corridors when we know pattern of dispersal. Intensive management of reserves once declared.
- Feral animal control: generally lower numbers of foxes and cats due to large dingo populations—research needed on this relationship and on direct predation by dingoes.
- Translocation: premature at this stage.
- Captive breeding: minimum population size studies. Captive stock for education of public. Captive animals used to develop radio collars for tracking in wild.
- Other: complete surveys in South Australia and Northern Territory (High priority). Perhaps use of local Aboriginal knowledge. Exclude visitation in some places. Prohibition of mineral exploration activities in prime habitat.

Organisations responsible for conservation of species: Queensland National Parks and Wildlife Service (QNPWS); South Australian National Parks and Wildlife Service (SANPWS); Conservation Commission of the Northern Territory (CCNT); Department of Primary Industry Queensland (QDPI).

Other organisation(s) or individual(s) involved: Australian National Parks and Wildlife Service (ANPWS). World Wide Fund for Nature Australia (WWF). Leong Lim.

Conservation reserves affected by recovery plan (list other action plan species in each conservation reserve): Innamincka Regional Reserve, South Australia.

Other public lands affected by action plan: (see *"Other public lands on which species occurs"* above). Species known range is all on public land at the moment.

Other lands affected by action plan: ?

Can recovery plan be carried out with existing resources? (If not, list additional staff and financial resources required): No.

1. Dedication of large reserve north from Birdsville.
2. Location of office plus full time ranger-naturalist in southwestern Queensland: Research—$A10,000 per annum plus costs and salary of research officer for two years—$A80,000 per annum for three years.

Perup Nature Reserve: *Myrmecobius fasciatus* (numbat), *Bettongia penicillata* (brush-tailed bettong), *Macropus eugenii* (Tammar wallaby), *Pseudocheirus occidentalis* (western ringtail possum). Lane Poole Reserve. Frank Hann National Park, Lake Magenta Nature Reserve, Kalbarri National Park (possibly).

Other public lands affected by conservation plan: State Forest in Conservation and Land Management (CALM's) Northern, Central, and Southern Forest Regions, vacant Crown land to east of wheatbelt.

Other land effected by conservation plan: Road reserves in wheatbelt.

Can recovery plan be carried out with existing resources? (If not, list additional staff and financial resources required): No.

- Effects of timber harvesting and prescribed burning in jarrah forest: and assessment of efforts of fox baiting; two staff for three years $A180,000 salaries, $A60,000 expenses.
- Development of translocation techniques: one staff for two years—$A60,000 salaries, $A25,000 expenses.

Total=$A325,000

Remarks: Species management program finalised, distribution data being collected and analyzed, surveys of peripheral areas commenced, chuditch awareness program underway. Monitoring of selected populations in jarrah forest ongoing. Recovery plan being prepared.

Species Recovery Outline No. 5

Family: Dasyuridae

Scientific name: *Parantechinus apicalis*

Common name: Dibbler

Intra-specific taxa: None

Species survival status: Endangered

Former distribution: Most of southwestern Western Australia. Recorded from the Moore River Region to King George Sound.

Current distribution: Restricted to two very small islands (Boullanger and Whitlock) off southwest Western Australian coastline and three widely separated mainland localities (see *"Conservation reserves on which species occurs"* below). Has suffered >90% decline in historically known range. Present distribution probably in preferred habitat.

Habitat: On mainland occurs now in dense heath habitat in coastal areas, with history of no recent burns.

Reasons for decline: Cat predation (speculative), fox predation (speculative), too frequent burning of habitat (speculative), habitat fragmentation (proven).

Conservation reserves on which species occurs: Boullanger and Whitlock Islands, Waychinicup National Park, Fitzgerald River National Park, Torndirrup National Park.

Other public lands on which species occurs: None known.

Other land on which species occurs: None known.

Is knowledge about species adequate for objectives and actions to be defined accurately? (If not, provide list of additional studies required): No.

- Effective monitoring (Fitzgerald River, Waychinicup, and Torndirrup National Parks) and genetic information (island populations) required.

Recovery plan objectives:

- Identify and remove factors causing historical decline in numbers.
- Monitor genetic diversity of island populations.
- Maintain current range and abundance.

Management actions completed or underway:

- All known populations in national parks and nature reserves, managed by Department of Conservation and Land Management.

Management actions required:

- Declaration of additional conservation reserves: no.
- Habitat management: management of preferred (current) habitats. Reduction of fire frequency in mainland areas of range, and maintenance of long-unburnt areas in major range. Control/manage all activities on islands. Maintain habitat free from *Phytophthora* (root rot fungi).
- Feral animal control: control of predators, especially foxes and cats, throughout the current range. Removal of mice from island populations.
- Translocation: not required at present.
- Captive breeding: ? Has not been bred in captivity.
- Other: population and habitat monitoring in all populations. Genetic monitoring in island populations. Population study of mainland populations to include radiotracking of mice from island populations. Local education programs. Biological surveys in southern-coastal Western Australia.

Organisations responsible for conservation of species: Western Australian Department of Conservation and Land Management (CALM).

Other organisation(s)-individual(s) involved: Chris Dickman, Sydney University. Garry Connell, University of Western Australia. Local communities.

Conservation reserves affected by recovery plan (list other action plan species in each conservation reserve): Boullanger and Whitlock Island Nature Reserves, Fitzgerald National Park, *Phascogale calura* (red-tailed phascogale), Waychinicup National Park, Torndirrup National Park.

Other public lands affected by conservation plan: None known.

Other lands affected by conservation plan: None known.

Can recovery plan be carried out with existing resources? (If not, list additional staff and financial resources required): No.

1. Population monitoring. If all known populations are censused four times a year, and genetic monitoring of island dibblers attempted every five years, costs in order of $A10,000-20,000 a year (x10 and not including salary cost of researchers).
2. Predator control. Unknown, but likely to involve high costs for both implementation and monitoring.
3. Mouse removal on islands probably achievable using selective pesticides for costs of <$A20,000.

Total (excluding predator control)=$A220,000+

Remarks: All known extant populations are protected (i.e. nature reserves and national parks).

Boullanger and Whitlock Islands harbour perhaps 200 animals. Torndirrup: unknown, probably <200. Other localities numbers unknown, though Jerdacuttup and Cheyne Beach populations possibly extinct.

Species Recovery Outline No. 6

Family: Dasyuridae

Scientific name: *Phascogale calura*

Common name: Red-tailed phascogale

Intra-specific taxa: None

Species survival status: Endangered (rare and restricted—drastic reduction in historical range).

Former distribution: Once widely but sparsely distributed across central and southern Australia, with populations in central-northern Western Australia, central Northern Territory, and the far western border between New South Wales and Victoria.

Current distribution: Semi-arid "wheatbelt" of southwest Western Australia. >90% decline in previous range.

Habitat: Confined to isolated reserves in the wheatbelt that receive an annual rainfall of between 300-600 mm. Prefers wandoo and rock oak communities.

Reasons for decline: Habitat fragmentation by clearing for agriculture (proven), predation by foxes and cats (speculative), changed fire regimes leading to very little old, long-unburnt vegetation (speculative).

Conservation reserves on which species occurs: East Yornaning, Tutanning Dongolocking, Parkeyerring Lake and Boyagin Nature Reserves, Fitzgerald River National Park. (Recorded in 1970s on Dragon Rocks, East Karlgarin and Bendering nature reserves.)

Other public lands on which species occurs: Dryandra and Highbury State Forest.

Other land on which species occurs: Private land near Tutanning Nature Reserve.

Is knowledge about species adequate for objectives and actions to be defined accurately? (If not, provide list of additional studies required): No.

- Additional population monitoring.
- Additional surveys needed.
- Effect of fox predation needs to be demonstrated.

Recovery plan objectives:

- Maintain current distributional range and abundance.
- Identify principal factors regulating the historical decline in range of populations (fox?, habitat?).
- Expand certain key reserves to enclose uncleared "preferred" habitat.

Management actions completed or underway:

- Possible benefit from fox control activities in Tutanning and Boyagin Nature Reserves and Dryandra State Forest.
- Fire management in Tutanning Nature Reserve and Dryandra State Forest.
- Study of species in progress (G.R. Friend & J.A. Friend).

Management actions required:

- Declaration of additional conservation reserves: yes, parts of Dryandra State Forest should be converted to National Park and Nature Reserve.
- Habitat management: careful management of preferred habitat, *Allocasuarina huegeliana—Eucalyptus wandoo* association in which this species is most abundant.
- Feral animal control: yes, if effect of fox predation is shown. Red-tailed phascogales are sensitive to 1080, but baits can be designed to prevent deleterious effects. National fox control program needed.
- Translocation: a future option, once suitable habitat is identified and secured as reserves.
- Captive breeding: not required at present.
- Other: surveys are needed to determine more accurately the species' abundance and distribution, and to gather more data on its specific habitat requirements. If suitable areas are found, they need to be incorporated into the reserve system urgently. Regular annual check of red-tailed phascogale population and habitat in key reserves (March-June trapping suggested). Research into species inter-relationship with fire and fire regimes on present reserves needs to be carried out (this is being partly covered with current work on Tutanning Nature Reserve, but a more intensive and widespread program requires initiation and funding). Education programs also need to be initiated.

Organisations responsible for conservation of species: Western Australian Department of Conservation and Land Management (CALM).

Other organisation(s) or individual(s) involved: Australian National Parks and Wildlife Service (ANPWS).

Conservation reserves affected by recovery plan (list other plan species in each conservation reserve): Yornaning Nature Reserve. Tutanning and Boyagin Nature Reserves: *Bettongia penicillata* (brush-tailed bettong), *Myrmecobius fasciatus* (numbat), *Macropus eugenii* (Tammar wallaby). Fitzgerald River National Park: *Parantechinus apicalis* (dibbler). Development of biological control agent for fox control would benefit all medium-sized mammals in southwest Western Australia.

Other public lands affected by conservation plan: Dryandra State Forest: *Myrmecobius fasciatus* (numbat), *Bettongia penicillata* (brush-tailed bettong), *Macropus eugenii* (Tammar wallaby).

Other lands affected by conservation plan: ?

Can recovery plan be carried out with existing resources? (If not, list additional staff and financial resources required): No.

1. Effect of fox control: costs $A20,000
2. Fire effects studies, 1 staff 3 years ($A90,000) costs ($A60,000)
3. Ongoing monitoring + new fox control: $A10,000 per annum (x10)

Total=$A270,000

Remarks: Survey may identify important populations which require fox control until biological control agent becomes available.

Species Recovery Outline No. 7

Family: Dasyuridae

Scientific name: *Sminthopsis douglasi*

Common name: Julia Creek dunnart

Intra-specific taxa: None.

Species survival status: Endangered (more information is needed. Only four specimens have been collected between 1931 and 1972. However, it comes from an area where there has been little biological survey carried out. The apparent rarity may be misleading).

Former distribution: No information.

Current distribution: Known from near Julia Creek and Richmond, northern Queensland.

Habitat: Unknown. The region in which it occurs has extensive natural grasslands and savannah woodlands with a rainfall of 450 mm.

Reasons for decline: Unknown. It is not known whether a decline has occurred.

Conservation reserves on which species occurs: None.

Other public lands on which species occurs: None

Other land on which species occurs: Recorded from Julia Creek, Garomna Station near Julia Creek, and Wyangerie Station near Richmond.

Is knowledge about species adequate for objectives and actions to be defined accurately? (If not, provide list of additional studies required): No.

- The prime need is for more survey in the region to determine its status and habitats.

Recovery plan objectives:

- Determine the status of the species.
- Determine some basic biology, particularly the habitat utilization.

Management actions completed or underway: None.

Management actions required:

- Declaration of additional conservation reserves: ?
- Habitat management: ?
- Feral animal control: ?
- Translocation: ?
- Captive breeding: ?
- Other: survey of distribution and abundance.

Organisations responsible for conservation of species: Queensland National Parks and Wildlife Service (QNPWS).

Other organisation(s) or individual(s) involved: Pat Woolley, La Trobe University.

Conservation reserves affected by recovery plan (list other action plan species in each conservation reserve): None.

Other public lands affected by conservation plan: None

Other lands affected by conservation plan: Possibly Garomna Station near Julia Creek and Wyangerie Station, near Richmond Queensland.

Can recovery plan be carried out with existing resources? (If not, list additional staff and financial resources required): No.

1. Funding for survey work is required. This might involve a grant for a few thousand dollars to a university or government researcher or a salary for about 1 year with associated field costs, e.g. $A50,000, depending on how work was organised.

Total=$A50,000

Remarks: This is a species that may not have declined and may or may not be rare. The area in which it occurs has received little biological survey and lack of records may merely reflect this. Other species of *Sminthopsis* manage to hang on in the Queensland pastoral areas and it is possible that this species is still present. The major requirement is for more survey work.

Species Recovery Outline No. 8

Family: Dasyuridae

Scientific name: *Sminthopsis psammophila*

Common names: Sandhill dunnart, Large desert sminthopsis

Intra-specific taxa: None

Species survival status: Vulnerable (rare throughout its range).

Former distribution: South Australia—Mamblyn and Boonerdoo on Eyre Penninsula and three localities in the Yellabinna sand dunes. Northern Territory—Lake Amadeus; Western Australia—southwestern corner of the Great Victoria Desert in and adjacent to the Queen Victoria Spring Nature Reserve.

Current distribution: As above, although not recorded on Eyre Peninsula since 1969, nor in the Northern Territory since 1895.

Habitat: Low parallel sandridges/hummock grass/open mallee-broombrush scrub/groves of desert oak (Aitken, 1983), see also Pearson and Robinson (1990)—"diverse shrub layer and a species of spinifex was always present with a cover of >20%".

Reasons for decline (state whether known or speculative): Not known whether any decline.

Conservation reserves on which species occurs: Queen Victoria Spring Nature Reserve, Yellabinna Regional Reserve.

Other public lands on which species occurs: None known.

Other land on which species occurs: Maralinga Tjarutja Aboriginal Lands; mining leases and vacant crown land in Western Australia.

Is knowledge about species adequate for objectives and actions to be defined accurately? (If not, provide list of additional studies required): No.

- Begin a full-time ecological study of the species both in the Queen Victoria Spring Nature Reserve, Western Australia and at the Yellabinna sites in South Australia.

- Conduct a systematic search in the area of Hambidge Conservation Park the nearest remaining natural vegetation to the rediscovery sites at Mamblyn and Boonerdoo.
- Surveys in Northern Territory and other suitable areas of the Great Victoria Desert.

Recovery plan objectives:
- To define the species distribution.
- To understand the habitat preferences of the species.
- To develop a conservation management plan for the species.

Management actions completed or underway:
- Occasional captures at a study site in Queen Victoria Spring (10 individuals, March 1986-January 1991).

Management actions required: Insufficient data available for management actions to be defined.
- Declaration of additional conservation reserves:?
- Habitat management: ?
- Feral animal control: ?
- Translocation: ?
- Captive breeding: ?
- Other: ?

Organisations responsible for conservation of species: South Australian National Parks and Wildlife Service (SANPWS); Western Australian Department of Conservation and Land Management (CALM).

Other organisation(s) or individual(s) involved: None known.

Conservation reserves affected by recovery plan (list other action plan species in each conservation reserve): Queen Victoria Spring Nature Reserve, Western Australia: *Dasycercus cristicauda* (mulgara), Yellabina Regional Reserve and Hambidge Conservation Park, South Australia.

Other public lands affected by conservation plan: ?

Other land affected by conservation plan: ?

Can recovery plan be carried out with existing resources? (If not, list additional staff and financial resources required): No.
1. Begin full-time ecological study of the species both in the Queen Victoria Spring Nature Reserve, Western Australia, and at the Yellabinna sites in South Australia and;
2. Conduct a systematic search in the area of Hambidge Conservation Park the nearest remaining natural vegetation to the rediscovery sites at Mamblyn and Boonerdoo: Two staff for three years—$A200,000 salaries, $A100,000 support.
3. Conduct surveys in the Northern Territory: Two staff for six months—$A50,000.

Total=$A350,000

Remarks: Very difficult species because of remoteness of habitat, apparent low densities and perhaps poor trapability.

Species Recovery Outline No. 9

Family: Peramelidae

Scientific name: *Isoodon auratus*

Common names: Golden bandicoot, Wintarru

Intra-specific taxa: *I.a. auratas* (mainland), *I.a. barrowensis* (Barrow Island).

Species survival status: Endangered

Former distribution: Widespread in arid deserts and adjacent semi-arid areas; distribution included Tanami, Gibson, Great Victoria, Great Sandy and Little Sandy Deserts; also in Kimberley of Western Australia and northern parts of the Northern Territory. Also on Barrow, Middle, Hermite (Pilbara), and Augustus Islands (Kimberley), Western Australia.

Current distribution: Has become extinct in almost all of its mainland range except a small area of the northwest Kimberley. Also extinct on Hermite Island. Remains on Barrow, Middle, and Augustus Islands.

Habitat: Spinifex and tussock grasslands.

Reasons for decline: Changed fire regimes (speculative), exotic predators, especially the European red fox (speculative), competition from rabbits (speculative).

Conservation reserves on which species occurs: Prince Regent Nature Reserve (Kimberley, Western Australia), Barrow Island and Middle Island Nature Reserves, Western Australia.

Other public lands on which species occurs: Augustus Island (Aboriginal Reserve).

Other land on which species occurs: None known.

Is knowledge about species adequate for objectives and actions to be defined accurately? (If not, provide list of additional studies required): No.
- Study ecology on Barrow Island.
- Current mainland range requires delimiting and mainland status needs defining.
- Study of general ecology and fire ecology and effects of feral cats on Kimberley population.
- Reasons for decline need definition (could be done as part of re-introduction program).

Recovery Plan objectives:
- Define present mainland range.
- Maintain present range.
- Set up captive breeding colonies from mainland populations.
- Reintroduce to areas of former range.

Management actions completed or underway:
- Pilot study conducted in 1988 to look at problems of capture, transport and maintenance in captivity of mainland population.
- Eradication of *Rattus rattus* on Middle Island to be carried out in 1991.
- Study of ecology on Barrow Island commenced 1990.

Management actions required:
- Declaration of additional conservation reserves: yes, but range needs definition first. Additional reserves are required in areas of former range in Northern Territory for reintroduction.
- Habitat management: fire management probably required.
- Feral animal control: foxes, feral cats, and *Rattus rattus* will need control if reintroduction is attempted.

- Translocation: should be reintroduced to parts of former range, especially deserts and Hermite Island.
- Captive breeding: yes, to produce mainland population animals for reintroduction.
- Other: examine genetic, physiological, and behavioural differences between Barrow Island and Kimberley populations.

Organisations responsible for conservation of species: Western Australian Department of Conservation and Land Management (CALM); Conservation Commission of the Northern Territory (CCNT) if species reintroduced to central Australia.

Other organisation(s) or individual(s) involved: West Australian Petroleum Pty. Ltd.

Conservation reserves affected by recovery plan. (List other action plan species in each conservation reserve): Prince Regent Nature Reserve. Barrow Island Nature Reserve: *Petrogale lateralis* (black-footed rock-wallaby), *Bettongia lesueur* (burrowing bettong).

Other public lands affected by conservation plan: Augustus Island.

Other land affected by conservation plan: None known.

Can recovery plan be carried out with existing resources?: No.

1. Definition of current range and setting up of captive breeding colony of mainland individuals: Two staff for two years—$A120,000 salaries, plus support including helicopter charter $A100,000.
2. Study of effects of fire and cats on Kimberley population: Two staff for three years—$A180,000 salaries, support including helicopter charter—$A150,000.
3. Reintroduction program: Two staff for five years—$A300,000 salaries, support $A170,000.
4. Ongoing predator control in re-introduction areas $A25,000 per annum (x10).

Total=$A1,170,000

Remarks: Some research can be integrated with reintroduction program.

Species Recovery Outline No. 10

Family: Perameloidea

Scientific name: *Perameles bougainville*

Common name: Western barred bandicoot

Intra-specific taxa: *P.b. bougainville* (Shark Bay islands), *P.b. fasciata*, (Victoria, New South Wales—extinct), *P.b. myosura* (Southwest Australia—extinct), *P.b. notina* (Western Australia, South Australia—extinct)

Species survival status: Endangered

Former distribution: From near Onslow in Western Australia southwards to near Perth and eastwards through southern South Australia to western New South Wales and northwestern Victoria; Bernier and Dorre Islands, Shark Bay.

Current distribution: Bernier and Dorre Islands.

Habitat: Semi-arid areas with a variety of vegetation types, including dense scrub, open bluebush and saltbush plains, and stony hills. On Bernier and Dorre Islands it is particularly common in sandhills behind beaches.

Reasons for decline: Predation by foxes and feral cats (speculative), competition from rabbits (speculative), habitat clearing and degradation.

Conservation reserves on which species occurs: Bernier and Dorre Islands Nature Reserve

Other public lands on which species occurs: None.

Other land on which species occurs: None.

Is knowledge about species adequate for objectives and actions to be defined accurately? (If not, provide list of additional studies required): No.

- Studies on biology on Dorre Island underway.

Recovery plan objectives:

- Maintain populations on Bernier and Dorre Islands.
- Reintroduce to areas of former range.

Management actions completed or underway:

- Reservation of all remaining habitat in nature reserves.
- Reservation of potential reintroduction areas on Shark Bay peninsulas.

Management actions required:

- Declaration of additional conservation reserves: not required at present.
- Habitat management: protection of Bernier and Dorre Islands from introduction of predators and frequent fire. Fencing of Shark Bay peninsulas and eradication of feral animals.
- Feral animal control: not required on Bernier and Dorre islands. Will be required if translocation to mainland or Dirk Hartog Island takes place.
- Translocation: translocation to Dirk Hartog Island (cat control required), and to Shark Bay peninsulas.
- Captive breeding: not required for species preservation/may be required if translocation proceeds.

Organisations responsible for conservation of species: Western Australian Department of Conservation and Land Management (CALM).

Other organisation(s) or individual(s) involved: None known.

Conservation reserves affected by recovery plan (list other action plan species in each conservation reserve): Bernier and Dorre Islands Nature Reserve: *Bettongia lesueur* (burrowing bettong), *Lagorchestes hirsutus* (mala), *Lagorchestes fasciatus* (banded hare-wallaby).

Other public lands affected by conservation plan: None.

Other lands affected by conservation plan: None.

Can recovery plan be carried out with existing resources?: No.

1. Stationing of ranger on Bernier and Dorre Islands: $A60,000 per annum (x10). (Also included in plans for *Bettongia lesueur* (burrowing bettong), *Lagorchestes hirsutus* (mala), and *Lagostrophus fasciatus* (banded hare-wallaby)).
2. Reintroduction to Dirk Hartog Island, including cat control: Two staff for four years—$A240,000 salaries, $A130,000 support. (Also

included in plans for *Bettongia lesueur* (burrowing bettong) and *Lagostrophus fasciatus* (banded hare-wallaby).

3. Reintroduction to one Shark Bay peninsula; 5 km fencing $A200,000, feral animal eradication and reintroduction, $A240,000 salaries, $A130,000 support.

Total=$A1,540,000

Species Recovery Outline No. 11

Family: Peramelidae

Scientific name: *Perameles gunnii*

Common name: Eastern barred bandicoot

Intra-specific taxa: None

Species survival status: Vulnerable (endangered in Victoria; vulnerable in Tasmania).

Former distribution: Formerly an extensive western Victoria distribution, also occurred in southeast South Australia; northern and eastern Tasmania.

Current distribution: Victoria (drastically reduced range), Hamilton only. Tasmania, as above with some extension of range into northwest and northeast areas, and some likely loss of range in Midlands.

Habitat: Open grasslands and grassy woodlands with available shelter. Suburban gardens and nearby grasslands (Hamilton). Tasmania, grasslands, grassy open forests and woodlands as well as some agricultural areas.

Reasons for decline: Loss of habitat, predation by cats and fox (proven), disease (speculative), and road kills (proven at Hamilton).

Conservation reserves on which species occurs: Victoria: Gellibrand Hill Park, "Mooramong", Skipton (National Trust property), Hamilton Community Parklands—breeding colonies as precursors to free-ranging colonies; Werribee Zoological Park, Cobra Killuc State Wildlife Reserve and Serendip (Lara) captive breeding colonies proposed. Tasmania; Lime Bay Nature Reserve, Mt. Field National Park (introduced at this location), Maria Island National Park, Asbestos Range National Park, Mt. William National Park. Habitat is poorly conserved by State Reserve system in Tasmania.

Other public lands on which species occurs: Small areas of public land (not necessarily state) at Hamilton (Victoria) only.

Other land on which species occurs: Victoria. Private land in city and adjacent Shire of Dundas. Tasmania, occurs widely on private land.

Is knowledge about species adequate for objectives and actions to be defined accurately? (If not, provide list of additional studies required): No.

- Still need information about dispersal and survival of young.
- Effects of pesticides.
- Tasmanian status and habitat association.
- Impacts of land management practices.
- Release techniques for re-establishment program.
- More specific habitat selection information at 'natural' sites.

Recovery plan objectives:

- Stabilise and increase population at Hamilton.
- Establish satellite breeding colonies.
- Establish satellite free-living colonies.
- Determine status of species in Tasmania.
- Determine taxonomic status of Tasmanian population.
- Study ecology of Tasmanian populations.

Management actions completed or underway: Victoria only.

- Victoria Department of Conservation and Environment (VDCE) has published a comprehensive management plan setting out management actions; Arthur Rylah Institute for Environmental Research, Technical Report Series No. 63, Peter R. Brown, February, 1989.
- Establishment of recovery teams.
- Stabilization and enhancement of local habitat.
- Reduce predation (fox and cat); motorcars are proven killers.
- Captive breeding
  - -Local reintroduction.
  - -Satellite colony development.
- Genetic management.
- Genetic research.
- Ecological research.
- Taxonomic research.

Management actions required: Victoria only.

- Declaration of additional conservation reserves: not really feasible. However, some areas of public land ('commons') are scheduled for future sites. They would then have a change of status. Will be attempting to establish, or re-establish colonies on existing ones.
- Habitat management: retention of remnant habitats on both public and private lands. Establishment of corridors. Ecosystem, habitat regeneration. Fencing of critical habitat. Increased management within reserves. Prohibition of habitat clearing. Deployment of management protection zones.
- Feral animal control: exclusion of domestic grazing (sheep). Upgrade existing legislation for cat control. Enforcement of laws on fox control. Predator control/biological research.
- Translocation: researching reintroduction techniques.
- Captive breeding: yes, but still need to improve survival of young. Also methods of locating/encountering weanlings.
- Other: protection from pesticides and cars. Use of media and community conservation organisations. Specific fundraising ventures. Multi-agency cooperation. Short- and long-term monitoring programs. Surveys of species dispersal areas.

Organisations responsible for conservation of species: Victorian Department of Conservation and Environment (VDCE). Portland, Ballarat, Melbourne, Geelong, Colac Regions Flora and Fauna Division. Tasmanian Department of Parks, Wildlife, and Heritage (TDPW&H).

Other organisation(s) or individual(s) involved: Victoria College (Peter R. Brown). LaTrobe University (Neil Murray). University of Melbourne. Zoological Board of Victoria (Graham Mitchell). Chicago Zoological Society. Local community at Hamilton. Museum of Victoria. National Trust. University of Tasmania (A. Bradley & R. Rose).

Conservation reserves affected by recovery plan (list other action plan species in each conservation reserve): (see *"Conservation reserves on which species occurs"* above).

Other public lands affected by conservation plan: Small areas of public land at Hamilton, Victoria.

Other land affected by conservation plan: Large areas of private land in Tasmania. Private land in City and adjacent Shire of Dundas in Victoria.

Can recovery plan be carried out with existing resources? (If not, list additional staff and financial resources required): No.

1. Estimates by Victorian government of ca. $A350,000 for 1991-1992, unlikely to be met. Additional funds essential for research, management and monitoring. Need to examine pesticide levels in both Victorian and Tasmanian populations. $A200,000-300,000 per annum for three to five years.
2. In Tasmania, research is needed to compare populations from 'natural' grasslands and grassy woodlands with those from agricultural areas which now comprise much of the species' current range. Two researchers for three years—$A60,000 salaries and $A25,000 support costs per annum.
3. Additional monies for Tasmanian status review $A35,000.

Total=$A1,790,000

Remarks: All management actions identified in Departmental Action Plan are addressed in the Final Management Plan (Victoria).

Species Recovery Outline No. 12

Family: Thalacomyidae

Scientific name: *Macrotis lagotis*

Common names: Bilby, Dalgyte, Ninu

Intra-specific taxa: *M.l. lagotis* (central Australia), *M.l. sagitta* (southwest Queensland), *M.l. cambrica* (New South Wales—extinct), *M.l. nigripes* (South Australia—extinct), *M.l. grandis* (South Australia—extinct), *M.l. interjecta* (southern Western Australia—extinct).

Species survival status: Vulnerable (range severely reduced. Endangered in Queensland. Distribution shown to have declined between 1970-1985. Extinct in southern half of range. Rare to uncommon, declining in numbers).

Former distribution: The arid and semi-arid regions throughout most of the Australian mainland south of about latitude 18° south and 25° to 26° in the Gibson Desert.

Current distribution: Currently restricted to parts of the Great Sandy, Gibson, and Tanami Deserts and the Pilbara with outlying population in southwest Queensland (Diamantina Shire), also west Kimberley (including Beagle Bay area). Restricted to low rainfall 200-250 mm desert areas. Sub-populations are distributed extremely patchily within the above range.

Habitat: Former habitat included savannah woodland/shrub grassland communities on a range of soil types. Now restricted to grasslands/acacia shrublands with spinifex/tussock grass. In southwest Queensland bilbies occur in Mitchell grassland, not hummock grassland.

Reasons for decline: On a broad overview, current bilby distribution coincides with areas where foxes and rabbits are absent or in low abundance and where grazing of stock is absent or conducted in low intensity. Fire is suggested as important in creating suitable habitat and promoting the production of certain food resources. Habitat degradation from the effects of rabbits, feral or controlled stock, predation by introduced predators, and changes in the fire regime are suggested as causing the decline of the bilby. The importance of individual factors varies in different parts of the bilby's range.

Current primary threat in the southwest of Queensland is change in land use due to "property improvement".

Conservation reserves on which species occurs: Western Australia: Gibson Desert Nature Reserve, Rudall River National Park. Watarrka National Park in the Northern Territory (reintroduced population). Queensland—None.

Other public lands on species occur: Pastoral leases and vacant Crown land.

Other land on which species occurs: Aboriginal land in Northern Territory and Western Australia.

Is knowledge about species adequate for objectives and actions to be defined accurately? (If not, provide list of additional studies required): Yes, but the following problems still must be addressed:

- Need to continue monitoring of population.
- Establish levels of genetic variation apportioned within the wild and captive population.
- Establish mechanisms and limits which make habitat suitable for occupation: (a) effects of rainfall events and fire on vegetation structure and cover composition; (b) effects of rabbits and stock, rainfall and fire on food production, i.e. seed, bulb, invertebrates; (c) effects of predator species on the population and acceptable limits of predator activity.
- Establish whether direct control of predators is feasible and effective on specific target species, i.e. foxes.
- Establish the effects of rabbit control on predator activity and availability of bilby foods.
- Examine the interaction between predator species and determine whether a preferred predator species may be encouraged.

Recovery plan objectives:

Northern Territory

- Ensure the survival of the bilby within its current range.
- Increase the range and abundance of the bilby within and on the periphery of its current distribution through habitat management.
- Increase the security and range of the species through the process of reintroduction into suitable, formerly occupied habitat.
- Further clarify distribution.
- Further clarify factors responsible for decline in numbers and range.

Queensland

- Obtain a reservation of suitable bilby habitat in Queensland for the species' permanent protection.
- Provide a staff presence to carry out management plans.
- Maintain captive colonies of bilbies for the purpose of reintroduction programs.
- Continue research program, particularly with respect to effects of stock grazing and predators.

Western Australia

- Declare reserves in the Great Sandy Desert where bilby occurs, e.g. Percival Lakes.
- Provide staff to carry out management plans.

Management actions completed or underway:

Northern Territory

- A previous survey funded by WWF (Australia) identified two forms of habitat which are important for the bilby in its current range. Areas with these characteristics need to be identified in its current range. The hypothesis that these areas form core breeding and dispersal areas for bilby needs to be tested.

- A reintroduction program has been initiated at Watarrka (Kings Canyon) National Park with funding from WWF Australia and the Winifred Violet Scott Fund. Replicates of this program are required to test the procedures and response of animals to release.

Queensland

- All current bilby habitat is managed for pastoral purposes. No bilby management is present. A grant of $A32,000 from Delhi Petroleum is funding a distribution and research project into bilbies presently conducted by Queensland National Parks and Wildlife Service.

Western Australia

- (as per description of Northern Territory survey above).
- Some areas of bilby habitat in conservation reserves.

Management action required:

- Declaration of additional conservation reserves: yes, reservation and protection of areas where bilbies occur at high density, e.g. Percival Lakes in Western Australia and paleo-drainage areas in Tanami Desert, Northern Territory.
- Habitat management: remove man-made water resources (bore, tanks, dams which might support elevated predator levels), determine whether habitat requires fire management, develop techniques for burning over extensive areas.
- Feral animal control: discourage non-native species, remove or reduce stock from areas of permanent protection, development and implementation of effective rabbit control measures, development and implementation of effective predator control measures (especially for foxes). Monitor the activity of predators and rabbits in habitats occupied by the bilby.
- Translocation: reintroduction studies: Collect data on: (1) home range, predation, social organization, and dispersal; (2) feeding trials and habitat manipulation to encourage individuals to remain at high density and in a localised predator managed situation; (3) response of predators and rabbits to control measures.
- Captive breeding: continuation of captive breeding programs in Alice Springs and Western Plains Zoo.
- Other: population monitoring, basic natural history research, identify important habitat assemblages and monitor bilby activity, monitor several assemblages to gauge the effects of bilby density and movements in response to rainfall events, fire, and food availability.

Organisations responsible for conservation of species: Conservation Commission of the Northern Territory (CCNT); Western Australian Department of Land Management (CALM); Queensland National Parks and Wildlife Service (QNPWS); New South Wales National Parks and Wildlife Service (NSWNPWS). South Australian National Parks and Wildlife Service (SANPWS).

Other organisation(s) or individual(s) involved: Landowners. World Wide Fund for Nature Australia (WWF). Winifred Violet Scott Fund. Western Plains Zoo.

Conservation reserves affected by recovery plan (list other action plan species in each conservation reserve): Western Australia—Gibson Desert Nature Reserve, Rudall River National Park. Watarrka National Park in the Northern Territory (reintroduced population). Queensland—none.

Fire management within the bilby's current range would be beneficial to all arid zone native fauna. Methodology to control foxes, feral cats, and rabbits in the arid zone would benefit most native fauna in the 0.1-5.0 kg range. Methodology and procedures for reintroduction of the bilby would be applicable and benefit programs working with 0.1-5.0 kg native fauna species.

Other public lands affected by plan: Pastoral lease and vacant Crown Land.

Other lands affected by plan: Aboriginal land in the Northern Territory and Western Australia.

Can recovery plan be carried out with existing resources? (If not, list additional staff and financial resources required): in the long term, No.

Queensland

1. Initial costs $A147,000 plus continuing yearly costs of $A50,000 (x10)

Northern Territory

1. Short term $A10,000 for coordination between states to establish goals and database; $A3,000 to establish genetic variability in wild population; and $A80,000 in first year to map habitat areas suggested by WWF report + $A50,000 per annum for years 2 and 3 of three-year program.
2. Reintroduction program; maintenance of a captive breeding colony $A10,000 (yearly cost per captive population x 3); $A3,000 for coordination between programs; two staff for 3 years—$A210,000 salaries, $A90,000 support; one staff for 7 years of continuing program—$A280,000 salaries, $A140,000 support for translocation programs.

Western Australia

1. Management of desert reserves: staff and support $A100,000 per annum (x10)

Total=$A2,597,000

Remarks: The identification of suitable habitat for the bilby could be incorporated into surveys of the resource-rich and/or paleo-drainage systems of the arid zone. It has been proposed that these areas form key habitats for many of the arid zone fauna and warrant detailed investigation.

Reintroduction programs for the bilby could be incorporated or expanded to accommodate other species where reintroduction methods need to be tested. Economies of scale would exist if another species were introduced (i.e. bilby, burrowing bettong, mala etc.) into the same area. Infrastructure development, habitat monitoring, predator control, and monitoring could be shared at the one site, as could equipment and manpower.

Species Recovery Outline No. 13

Family: Vombatidae

Scientific name: *Lasiorhinus krefftii*

Common names: Northern hairy-nosed wombat, Yaminon

Intra-specific taxa: None

Species survival status: Endangered (range severely reduced, only one population of 65 individuals known).

Former distribution: Wellington Caves, New South Wales (1869); Jerilderie, New South Wales (1884); Moonie River, Bullamon Plains & Balonne (near St. George), Southern Queensland (between 1891 and 1900); and perhaps near Injune and Tambo in Queensland prior to 1900?

Current distribution: Central Queensland, Belyando River Catchment, Epping Forest National Park, near Clermont.

Habitat: Prefers acacia and eucalypt woodland combined with patches of closed scrub with native grasses as ground cover, and flat, sandy, and semi-arid grasslands. Lives in burrows, preferably on banks of previous watercourses.

Reasons for decline: Grazing competition with cattle, sheep, and rabbits, habitat destruction (not proven but strong circumstantial evidence).

Conservation reserves on which species occurs: Epping Forest National Park.

Other public lands on which species occurs: None.

Other land on which species occurs: None.

Is knowledge about species adequate for objectives and actions to be defined accurately? (If not, provide list of additional studies required): Yes, but following problems must still be addressed:

- Adequate monitoring of population size.
- Possible influence of reduced variability and inbreeding on population viability.
- Securing and rehabilitation of additional habitat for relocation.

Recovery plan objectives:

- Ensure increase in population to level sustainable by habitat in Epping Forest National Park.
- Establish populations in other areas formerly occupied by the species.

Management actions completed or underway:

- Sustainable population plan above is being pursued.
- Protection of occupied habitat from fire and cattle grazing, restriction of visitation.
- Study of behaviour, ecology, and genetics.
- Fencing off critical habitats.
- Fire regime research.

Management actions required:

- Declaration of additional conservation reserves: not yet required.
- Habitat management: continued protection from fire. Retention of remnant habitats on private and public land. Possible establishment of corridors and habitat regeneration. Possibly increased reserve management.
- Feral animal control: continued exclusion of cattle.
- Translocation: relocation of breeding animals to other suitable sites. Research into reintroduction techniques.
- Captive breeding: possibly in future (zoo technology for captive breeding needs reassessment).
- Other: continued monitoring of population size and breeding. Possible future control of kangaroos, dingoes. Research on climatic warming. Research on native animal grazing competition and predation (dingo). Research into the role of fires in the maintenance of grassland species and structural diversity.

Organisations responsible for conservation of species: Queensland National Parks and Wildlife Service (QNPWS).

Other organisation(s) or individual(s) involved: Queensland Department of Primary Industries (QDPI); Australian National Parks and Wildlife Service (ANPWS); La Trobe University Department of Genetics (Neil Murray); Chris Johnson (University of Tasmania).

Conservation reserves affected by recovery plan (list other action plan species in each conservation reserve): Epping Forest National Park and other reserved areas in Brigalow belt of Central Queensland.

Other public lands affected by conservation plan: None.

Other land affected by conservation plan: None.

Can recovery plan be carried out with existing resources? (If not, list additional staff and financial resources required): Yes.

Note: Additional funding may be required for relocation. Management costs from QNPWS resources. Additional funding sought to meet the monitoring guidelines, established by Dr. C.N. Johnson; recommended a re-trapping program once every three years, to establish wombat condition, breeding, and recruitment:

1. Monitoring. Not expensive, mostly incorporated with routine management exercise, but extra resources might be required for census by re-trapping every three years—$A3,000-4,000 (x3).
2. Relocation. This may look like a viable option in 5 to 10 years, and will be an expensive exercise, requiring a full-time wildlife biologist, field assistant, vehicle plus running expenses, fencing material for enclosures, and possibly construction of facilities for holding animals in captivity, radio-location equipment etc.—$A60,000 per annum salaries, $A30,000 for vehicle purchase, $A40,000 support. This program would need at least four years of intensive work, plus follow-up, at a total cost of approximately $A330,000.
3. If suitable sites currently exist, they need to be secured now. If it becomes necessary to relocate to altered habitats, it is equally important to acquire the land now to allow for rehabilitation to commence well before the relocation of animals: $A250,000 for land acquisition; $A15,000 salaries (6 months); $A10,000 costs for site identification and preparation of acquisition proposals.
4. Genetics. This is the outstanding aspect of the current research program—blood samples are being held by the marsupial conservation genetics group at La Trobe University— $A25,000.

Total=$A642,000

Remarks: All habitat known to be occupied by the northern hairy-nosed wombat is now reserved in Epping Forest National Park. In response to preliminary research results, the park is kept free of cattle and fire. A detailed research program, examining wombat behaviour, ecology, genetics, and pasture ecology is nearing completion. Results will be applied to long-term management of the population and habitat relocation has been deferred until the population expands to occupy all habitat available in the park. Research shows that almost all females are breeding and young are being recruited into the adult population.

Species Recovery Outline No 14

Family: Potoroidae

Scientific name: *Bettongia lesueur*

Common names: Boodie, Burrowing bettong

Intra-specific taxa: None

Species survival status: Endangered

Former distribution: Widespread across semi-arid and arid Australia and areas of higher rainfall with Mediterranean climate. Not recorded in Queensland. Barrow, Boodie, Bernier, Dorre, and Dirk Hartog Islands.

Current distribution: Barrow, Bernier, and Dorre Islands. Extinct on mainland, on Dirk Hartog Island (1920s) and on Boodie Island (1988).

Habitat: Very broad range of habitats from spinifex deserts to woodlands. Lives in warrens.

Reasons for decline: Predation by foxes (speculative), predation by cats, particularly on islands (speculative), competition from rabbits (speculative), changed fire regimes (speculative), competition from black rats on Boodie Island (speculative).

Conservation reserves on which species occurs: Barrow Island Nature Reserve, Bernier and Dorre Islands Nature Reserves.

Other public lands on which species occurs: None.

Other land on which species occurs: None.

Is knowledge about species adequate for objectives and actions to be defined accurately? (If not, provide list of additional studies required): No.

- Investigation of differences between island populations required.
- Research into food, interactions with foxes, cats, rabbits, and fire needed.

Recovery plan objectives:

- Retain current range and abundance.
- Study interactions with foxes, cats, rabbits, and fire regimes to identify causes of historic decline.
- Reintroduce to areas of former range, including islands.

Management actions completed or underway:

- Reservation of all remaining habitat.
- Survey of distribution, abundance and habitat requirements on Barrow, Boodie, Bernier, and Dorre Islands.
- Translocation experiments to Heirisson Prong (Shark Bay) and Gibson Desert being developed.

Management actions required:

- Declaration of additional conservation reserves: yes—there is no unreserved habitat within current range. Declaration of peninsulas on southern shores of Shark Bay (Peron, Heirisson, Bellefin, and Steep Point) would enhance prospects for reintroduction.
- Habitat management: not required, but may be necessary if translocation to mainland takes place.
- Feral animal control: not required for existing island populations; will be required if translocation occurs.
- Translocation: yes, to Dirk Hartog Island (cat control necessary), proposed national parks and other areas in Shark Bay (fencing and feral animal control necessary) and to Gibson Desert Nature Reserve.
- Captive breeding: not required for conservation of species. May be necessary for translocation.

Organisations responsible for conservation of species: Western Australian Department of Conservation and Land Management (CALM).

Other organisation(s) or individual(s) involved: CSIRO Division of Wildlife and Ecology; West Australian Petroleum Pty. Ltd.; World Wide Fund for Nature Australia (WWF); Australian National Parks and Wildlife Service (ANPWS).

Conservation reserves affected by recovery plan. (List other action plan species in each conservation reserve): Barrow Island Nature Reserve: *Petrogale lateralis* (black-footed rock-wallaby), *Isoodon auratis* (golden bandicoot). Bernier and Dorre Islands Nature Reserve: *Perameles bougainville* (western barred bandicoot), *Lagorchestes hirsutus* (mala), *Lagostrophus fasciatus* (banded hare-wallaby).

Other public lands affected by recovery plan: Dirk Hartog Island. Heirisson Prong, Shark Bay. Bellefin Prong, Shark Bay. Peron Peninsula, Shark Bay. Gibson Desert Nature Reserve: *Macrotis lagotis* (bilby), *Dasycercus cristicauda* (mulgara).

Can recovery plan be carried out with existing resources?: No.

1. Reintroduction to Dirk Hartog Island, including cat control: Two staff for four years—$A240,000 salaries, $A130,000 support. (Also included in plan for banded hare-wallaby, western barred bandicoot).
2. Stationing of ranger on Bernier and Dorre Islands: $A60,000 per annum (x10) (also for banded hare-wallaby, mala, western barred bandicoot).
3. Reintroduction to Shark Bay peninsulas. Construction of fence across Heirisson Prong carried out by Useless Loop community. Will require fox and rabbit control—$A10,000 per annum (x10), translocation of animals from Bernier or Dorre Islands—$A3,000 and monitoring of population—$A25,000 per annum (x10). The use of other peninsulas (Peron, Bellefin, and Steep Point) would require additional funding. [*Note:* 12 bettongs were successfully transferred to Heirisson Prong on May 7, 1992 (J. Short, pers. comm.)].
4. Reintroduction to Gibson Desert Nature Reserve: Two staff for five years—$A300,000 salaries, support—$A170,000 (in conjunction with reintroduction of the golden bandicoot).

Total=$A1,793,000

Remarks: Current study into ecology of island populations being carried out by CSIRO Division of Wildlife and Ecology, with support from Australian National Parks and Wildlife Service, World Wide Fund for Nature Australia and West Australian Petroleum Pty. Ltd.

Species Recovery Outline No. 15

Family: Potoroidae

Scientific name: *Bettongia penicillata*

Common name: Brush-tailed bettong

Intra-specific taxa: *B.p. ogilbyi* (southern and southwestern Australia), *B.p. penicillata* (southeastern Australia, extinct)

Species survival status: Endangered (rare, very restricted range [Western Australia] reintroduced and introduced island populations in South Australia).

Former distribution: From Shark Bay in Western Australia south and east covering all of southwest Western Australia, most of Southern Australia, but excluding the far southeast and northeast portions of the state, the northwest corner of Victoria and across the central portion of

distribution lies within production forest. Forest is available for harvesting, not all allocated for such. Known only from a single location in New South Wales. This isolated site in the Rockton area of Bondi State Forest is estimated to embrace approximately 10 sq. km. This area also lies within production forest. [*Note:* The species presence in this area has now been confirmed on several subsequent occasions.]

Habitat: Currently considered to be restricted to riparian vegetation with dense understoreys surrounded by mixed-species eucalypt forest. In the lowlands, generally found in remnant warm temperate patches or wet sclerophyll forests with rainforest elements. May be found in broader riparian habitats such as fern gullies or *Gahnia*-dominated herbfield in montain areas.

Reasons for decline: Now probably restricted to dense understoreys due to feral predator (fox and dog) pressure (speculative). Such pressure is being increased as a result of forest management practices, which facilitate ingress into reparian areas, e.g. roading, fuel reduction burning, clear-felling, etc.

Clear-felling and slash burning of mixed forests surrounding habitats are speculated to lead to local extinctions due to habitat depletion, changes in hydrology leading to reduction of fungal food resources, and long-term changes in rainforest floristics and structure.

Intense and broad scale wildfires can result in the permanent devastation of remnant rainforests with the subsequent loss of resident colonies. The probability of this occurring is increasing as areas of single species regrowth are expanded.

Additional concerns are the insularization of colonies. There is strong evidence to suggest that the isolation of small colonies will lead to their extinction. Sufficient contingency and linking habitat must be provided to allow migration between patches.

Impending integrated harvesting (Value Added Utilization System—VAUS) is likely to expand the above-mentioned problems in Victoria. Existing woodchip practices threaten the NSW colony and other potential habitats.

Conservation reserves on which species occurs: Rodger Wilderness Area, Snowy River National Park and possibly Errinundra National Park in Victoria. It should be noted that occurrence in reserves listed in Victoria is based on few data. Populations are *not* readily encountered and presence/absence is very difficult to confirm and quantify. Should not consider the reserves adequate. New South Wales—none.

Other public lands on which species occurs: State Forests in Victoria and New South Wales (Sheep Station Creek, Bondi State Forest).

Other lands on which species occurs: None known.

Is knowledge about species adequate for objectives and actions to be defined accurately? (If not, provide list of additional studies required): No

- Little or nothing known of the ecological requirements of the species' major food resource: hypogeal fungi and how they are affected by management (study currently in progress at ANU).
- Need to identify accurately location, size of populations in New South Wales; also habitat assessment.
- Distribution, status, habitat/diet needs/availability, other ecological parameters, reproductive physiology; most important population survey is movement dispersal, etc., probably to be undertaken in 1991-1992.

Recovery plan objectives:

- Determine current range, status, diet/food resource, reproduction, and habitat requirements.
- Maintain current range and abundance.
- Determine effects of forest management activities.
- Determine effects of habitat manipulation.
- Formulation of management strategy.

Management actions completed or underway:

- Victorian Department of Conservation and Environment (VDCE) Australian National Parks and Wildlife Service (ANPSW) have published detailed preliminary recommendations for management of *Potorous longipes* in Victoria (Scotts & Seeback 1989).
- Research funded jointly by VDCE and ANPWS completed in 1990 (field work completed, report in progress). New work is commencing in NSW in 1991.
- Species is to be listed under the Victorian "Flora and Fauna Guarantee Act".
- Management zone established in Victoria.
- In Victoria, the Potoroo Management Committee has refined prescriptions, and guidelines for management are being written.
- In New South Wales there is presently a moratorium on harvesting in the Sheep Station Creek area, although existing logpiles will be removed prior to closure of logging roads. The area is proposed to be included in the new Genoa National Park.
- Research is underway on the ecological role of hypogeal fungi, jointly funded by ANPWS and VDCE.
- NSW NPWS funded work by Australian National University (T. Norton and M. Saxon), on "Field Survey Design and Resource Requirements for Sampling Potential Habitat."

Management actions required:

- Declaration of additional conservation reserves: yes, reservation and protection of essential habitat when that is known. Protection of all known localities—Potoroo Protection Zones/Special Management Areas. Mechanisms already in place in Victoria for immediate protection of new colonies as they become known. Sheep Station Creek in New South Wales is to be listed on the Register of the National Estate and requires dedication as an appropriate reserve. Genoa National Park has been recommended but not yet gazetted.
- Habitat management: New South Wales—moratorium on logging on Sheep Station Creek. Additional and intensive surveys of all other areas predicted as potential potoroo habitat. Interim cessation of harvesting/road building activities in these areas. Possible management of habitat in younger regenerating forest habitats (not necessary in old-growth habitats). Habitat requirements to be investigated. Controlled fire regimes/research (experimental burning is proposed). Retention of remnant habitats on public land. Establishment of corridors and habitat rehabilitation. Increased management within reserves.
- Feral animal control: yes: fox and dog control in known population areas, coupled with predator interaction/biological studies (fox-dingo). Predator control will occur.
- Translocation: no.
- Captive breeding: yes, establishment of breeding colony at Melbourne Zoo or Taronga (colony currently exists at Healesville Sanctuary). Studies of reproductive physiology and behaviour are being done, with the aim to improve reproduction in captivity. An additional colony is to be set up when animals are available.
- Other: establishment of Management Committee (Recovery Team). Additional and continuing ecological/biological and abundance studies. Cessation of logging activities is necessary for the protection of all known populations. Education programs/use of media. Research on climatic warming on existing habitats. Long-term

monitoring programs (existing marked populations). Surveys of much of East Gippsland for further populations. Research into hypogeal fungi. Cooperation of timber industry. Genetic research.

Organisations responsible for conservation of species: Victorian Department of Conservation and Environment (VDCE): Flora and Fauna Division, Forests Division, Orbost Region; New South Wales National Parks and Wildlife Service (NSW NPWS); Forestry Commission of New South Wales.

Other organisation(s) or individual(s) involved: Zoological Board of Victoria; Australian National Parks and Wildlife Service (ANPWS); Australian National University (ANU).

Conservation reserves affected by recovery plan (list other action plan species in each conservation reserve): Rodger Wilderness Area, Snowy River National Park and Errinundra National Park in Victoria. All potential long-footed potoroo habitats.

Other public lands affected by plan: State Forests in Victoria and New South Wales (Sheep Station Creek—Bondi State Forest).

Other land affected by conservation plan: Probably not applicable.

Can recovery plan be carried out with existing resources? (If not, list additional staff and financial resources required): No.
Note: Australian National Parks and Wildlife Service has provided funds in Phase 1 and 2 of research project—total so far $A135,000. Victorian Dept. Conservation and Environment costs, $A50,000. This study is now nearing completion. Preliminary management recommendations are being adopted and implemented—zoning, forest management modification. Overall cost by end of 1989/1990 was $A300,000.

1. Additional surveys, general predator control, fire research, continued monitoring of the one accessible colony in Victoria. Predator control—three years, $A100,000 per annum. Program implementation and other ecological and survey costs, three years, $A100,000 per annum.
2. Intensive survey of sites in southeast New South Wales which occur in State Forest and investigate possibility of surveying sites on private land. Project leader and field team of two people for two years, $A130,000 per annum.

Total=$A860,000

Remarks: Resources are inadequate to service studies of the ecology of this species, its habitat, and food resources.

Of great concern is the rate of timber harvesting across the known range of the species. While draft prescriptions are currently being implemented, the Orbost Region of VDCE will be unable to effectively service the prescriptions at the existing staff levels and expertise.

Current surveys only have the resources to locate colonies on a broad scale. To facilitate effective and long-term conservation, we must aim to locate and manage as many clusters of colonies as possible, and do so within the time constraints of proposed timber harvesting rates. This is particularly so in areas earmarked for Value Added Utilization System trials.

Species Recovery Outline No. 18

Family: Macropodidae

Scientific name: *Lagorchestes hirsutus*

Common names: Mala, Rufous hare-wallaby

Intra-specific taxa: *L.h. hirsutus* (mainland subspecies), *L.h. bernieri* (Bernier Island, Western Australia), *L.h. dorreae* (Dorre Island, Western Australia)

Species survival status: Endangered

Former distribution: Formerly occupied >25% of continental Australia, covering most of the arid and semi-arid Australian zones. Western Australia—wheatbelt east to South Australian border north of Nullarbor Plain; north to Shark Bay and east through Great Sandy and Gibson Deserts to Northern Territory border. Northern Territory—spinifex deserts from Stuart Highway west to Western Australian border. South Australia—northwestern desert area.

Current distribution: Now only one small mainland population remaining in the Tanami Desert comprising approximately 30 individuals. Satisfactory numbers on Bernier and Dorre Islands. [*Note:* This population apparently wiped out by fire in November, 1991.]

Habitat: Arid and semi-arid parts of Australia, particularly spinifex hummock grasslands of the sand plain and sand dune deserts.

Reasons for decline: Changed fire regimes, fox predation (speculative), competition with rabbits (speculative), clearing and fragmentation of habitat in southwest Western Australia.

Conservation reserves on which species occurs: Bernier (4267 ha) and Dorre Islands (5163 ha) in Western Australia, both Class A Nature Reserves.

Other public lands on which species occurs: None.

Other land on which species occurs: Aboriginal freehold land in the Tanami Desert of the Northern Territory.

Is knowledge about species adequate for objectives and actions to be defined accurately? (If not, provide list of additional studies required): No.

- Methods for broad-scale fox control and/or reducing their impact.
- Appropriate fire management to improve and sustain suitable habitat.
- Techniques for reintroduction.
- Ecological research on the species, including processes of dispersion of individuals to colonise new areas, role of fire in enhancing or suppressing dispersion, and social organization as it effects recruitment and dispersion.
- Role of saline paleo-river systems as key refugia during periods of drought and appropriate management of these areas.
- Long-term effects of fire on habitat structure and composition and the role of fire intensity, season, specific habitat type, and preceding and subsequent rainfall on post-fire vegetation recovery.

Recovery plan objectives:

- Ensure survival of remaining wild populations
- Research factors related to the decline of the species.
- Establish zoo populations of the mainland subspecies to preserve gene pool.
- Extend current range by translocation of captive-bred animals to establish new wild populations.
- Implement appropriate management to encourage expansion of mainland populations.

Management actions completed or underway:

- Captive population established from recently extinct colony is breeding well and numbers 45 individuals, and approximately 60 are now part of a reintroduction program.
- Zoo populations established at Western Plains Zoo and Northern Territory Wildlife Park with proposal to provide: (1) Animals for reintroduction in the long term; (2) Access to scientists for study.
- Preliminary work being undertaken to investigate reintroduction techniques.

Management actions required:

- Declaration of additional conservation reserves: yes, Tamani Desert where remaining population exists and where species is being reintroduced.
- Habitat management: controlled fire regimes are an essential element of habitat management. Research required into interactions between fire intensity, season, preceding and subsequent rainfall, habitat type and resulting post-fire recovery of vegetation (medium priority). Exclude unnecessary disturbance by roadworks of remaining population by Tanami highway. Habitat regeneration through fire management. Fencing habitat not a preferred option, but may be relevant at a later stage. Establish appropriate fire regimes at the Tanami Desert population. Protect Bernier and Dorre Islands from frequent and extensive fires.
- Feral animal control: development of predator control program. Fox highest priority, feral cat also significant (native dingo significant in early phases of reintroduction). Investigate strategic predator control regarding seasonal conditions and abundance of alternate prey species. Role of rabbits during periods of stress (drought).
- Translocation: reintroduction to place where colony recently became extinct and to another area, also on Aboriginal land, where species is known to have occurred.
- Captive breeding: maintenance of captive breeding population to produce animals for reintroduction programs. Breed well in captivity. High potential to re-establish species in northern desert areas where foxes are rare or absent. Also relevant in southern areas where effective predator control maintained.
- Other: population monitoring. Agreements with Aboriginal land owners for conservation (negotiation with Warlpiri traditional owners for conservation management in the Tanami Desert). Investigation of genetic differences between the two mainland and two island gene pools. Continued publicity to increase public awareness of the critical status of desert mammals. Vast majority of former range held as Aboriginal Freehold Title or with strong Aboriginal affiliation. Aboriginal participation in program essential. Aboriginal knowledge and skills significant (also liaison regarding hunting activities and need to protect the mala). Television documentary of historical changes to the Australian fauna, the continued decline, and courses of action to reverse the situation. Clarify sub-specific differences. Ecological/biological surveys of eastern Great Sandy Desert Region. Exclude public visitation. Clarify role of fire with public and impact of predators.

Organisations responsible for conservation of species: Conservation Commission of the Northern Territory (CCNT). Western Australian Department of Conservation and Land Management (CALM). Central Land Council and Warlpiri Aboriginal owners and custodians.

Other organisation(s) or individual(s) involved: Australian Heritage Commission (AHC); Australian National Parks and Wildlife Service (ANPWS); World Wide Fund for Nature Australia (WWF); Western Plains Zoo; Northern Territory Wildlife Park.

Conservation reserves affected by recovery plan (list other action plan species in each conservation reserve): Bernier and Dorre Islands Nature Reserve: *Perameles bougainville* (western barred bandicoot), *Bettongia lesueur* (burrowing bettong), *Lagostrophus fasciatus* (banded hare-wallaby)

Other public lands affected by conservation plan: None

Other lands affected by conservation plan: Aboriginal freehold land in the Tanami Desert of the Northern Territory.

Can recovery plan be carried out with existing resources? (If not, list additional staff and financial resources required): No.

1. Reintroduction of captive bred animals into Tanami Desert, development of reintroduction techniques and post-reintroduction management, employment of Aboriginal traditional owners. One technical assistant for three years—$A90,000 salaries, $A45,000 support, $A100,000 automatic radio tracking system. Salary and support to extend tenure of consulting scientists for two years $A80,000 salaries, $A30,000 support.
2. Rangers for Bernier and Dorre Islands Nature Reserve—$A60,000 per annum (x10), (also included in outlines for burrowing bettong, banded hare-wallaby, and western barred bandicoot).
3. Long-term study of the role of fire intensity, season, pre- and post-fire rainfall and habitat type on structure and composition in post fire recovery of vegetation in hummock grasslands. Ten-year study at four months per annum. Two staff $A200,000 salaries, $A130,000 support.
4. Evaluate the role of key refugia habitats (saline paleo-river habitats; run-on areas) for persistence of mala during drought. Two staff for 3 years, $A210,000 salaries, $A90,000 support. Continued monitoring for 7 years at 4 months per annum, salaries $A170,000, support $A70,000.
5. Identification of genetic differences between island and mainland subspecies. Support for collection and analysis of tissue samples—$A15,000 including travel to islands and contracting for laboratory analysis.

Total=$A1,830,000

Remarks: Other management/research projects that would be desirable: Comparative population ecology of *L. conspicillatus* (spectacled hare-wallaby) and *O. unguifera* (northern nailtail wallaby) which are sympatric with *L. hirsutus* to improve knowledge of colonisation processes and response to stress in desert macropods.

Species Recovery Outline No. 19

Family: Macropodidae

Scientific name: *Lagostrophus fasciatus*

Common names: Banded hare-wallaby, Murning

Intra-specific taxa: *L.f. fasciatus* (Shark Bay Islands) *L.f. albipilis* (Southwest Australia—extinct)

Species survival status: Endangered

Former distribution: Semi-arid parts southwest Western Australia, Dirk Hartog, Bernier and Dorre Islands, Shark Bay.

Current distribution: Bernier and Dorre Islands, Western Australia.

Habitat: Scrub.

Reasons for decline: Competition from rabbits (speculative), predation by cats (speculative). Extinction on mainland may have occurred before arrival of fox in Western Australia.

Conservation reserves on which species occurs: Bernier and Dorre Islands Nature Reserve.

Other public lands on which species occurs: None.

Other land on which species occurs: None.

Is knowledge about species adequate for objectives and actions to be defined accurately? (If not, provide list of additional studies required): No.

- Study required of basic biology and effects of competition from rabbits.
- Techniques for eradication of feral cats from Dirk Hartog Island require development.

Recovery plan objectives:

- Maintain populations on Bernier and Dorre Islands.
- Reintroduce to Dirk Hartog Island.
- Reintroduce to mainland conservation reserves.

Management actions completed or underway:

- Protection of all current habitat in Bernier and Dorre Islands Nature Reserve.

Management actions required:

- Declaration of additional conservation reserves: yes, Dirk Hartog Island should be reserved and reintroduction carried out.
- Habitat management: protection of Bernier and Dorre Islands from frequent or extensive fire and from introduction of predators.
- Feral animal control: yes, cats and goats on Dirk Hartog Island and any other areas where reintroduction is contemplated.
- Translocation: to Dirk Hartog Island and to mainland conservation reserves.
- Captive breeding: yes, especially to provide animals for reintroduction.

Organisations responsible for conservation of species: Western Australian Department of Conservation and Land Management (CALM).

Other organisation(s) or individual(s) involved: World Wide Fund for Nature Australia (WWF). Perth Zoo may be involved at a later stage.

Conservation reserves affected by recovery plan (list other action plan species in each conservation reserve): Bernier and Dorre Islands Nature Reserve: *Perameles bougainville* (western barred bandicoot), *Bettongia lesueur* (burrowing bettong), *Lagorchestes hirsutus* (mala).

Other public lands affected by conservation plan: None.

Other lands affected by conservation plan: None.

Can recovery plan be carried out with existing resources?: No.

1. Stationing of ranger on Bernier and Dorre Islands—$A60,000 per annum (x10). (Also for mala, western barred bandicoot, burrowing bettong).
2. Study of basic biology and effects of rabbits. Two staff for two years—$A120,000 salaries, $A70,000 support.
3. Reintroduction to Dirk Hartog Island, including cat control. Two staff for four years—$A240,000 salaries, $A130,000 support. (Also included in plans for burrowing bettong and western barred bandicoot).

Total=$A1,160,000

Remarks: Reintroduction to Dirk Hartog Island was attempted in the 1970s, but was unsuccessful.

Species Recovery Outline No. 20

Family: Macropodidae

Scientific name: *Macropus eugenii*

Common name: Tammar wallaby

Intra-specific taxa: Populations from Western Australia and South Australia are quite distinct on basis of enzyme and protein polymorphisms. Populations within Western Australian grouping may be separated. Perup population shown to be genetically distinct. Populations in Western Australia and South Australia (Kangaroo Island) morphologically distinct (W. Poole, pers. comm).

Species survival status: Vulnerable

Former distribution: Semi-arid areas of the southwest of Western Australia and South Australia (from Ceduna across Eyre Peninsula) and the following islands: East Wallabi and West Wallabi (Houtman Abrolhos), Garden (near Perth), Middle and North Twin Peaks (Archipelago of the Recherche) Western Australia; St. Peter, Kangaroo, Flinders, and Thistle (South Australia). Also once common in high rainfall areas in the extreme southwest of Western Australia (W. Poole, pers. comm). Sub-fossil material from St. Francis, Reevseby, and Wedge Islands. Not known whether these represent natural populations at time of European settlement, or introductions by European sailors in 18th or early 19th century.

Current distribution: Remnant populations in southwest of Western Australia. Extinct on mainland of South Australia and on Flinders, St. Peter, and Thistle Islands. Abundant on Kangaroo Island. Introduced to Granite, Greenly, and Boston Islands, South Australia. Introduced to Kawau Island, New Zealand in 1870 and then to Rotorua circa 1912 (W. Poole, pers. comm).

Habitat: Dry sclerophyll forest, heathlands, coastal scrub, mallee, and woodland thickets.

Reasons for decline: Clearing of habitat (proven), predation by feral cats and foxes (foxes proven), changed fire regimes (proven), competition with introduced herbivores (sheep, cattle, rabbits).

Conservation reserves on which species occurs: Western Australia: Perup Nature Reserve, Tutanning Nature Reserve, Boyagin Nature Reserve, Kalbarri National Park, Houtman Abrolhos Nature Reserve, Archipelago of the Recherche Nature Reserve. South Australia: Greenly Island Conservation Park, plus all Kangaroo Island Parks: Beyeria, Cape Hart, Cape Gantheaume, Cape Torrens, Dudley, Kelly Hill, Lathami, Mount Taylor, Nepean Bay, Parndana, Pelican Lagoon, Seal Bay, Seddon, Vivonne Bay, and Western River Conservation Parks, Flinders Chase National Park.

Other public lands on which species occurs: Western Australia: Dryandra State Forest, Peniup Creek Reserve, Garden Island (Commonwealth Defence Land). South Australia: Granite Island (Local Government Reserve).

Other land on which species occurs: Private land near Cranbrook Western Australia, private land on Kangaroo Island (including many areas protected by Heritage Agreements) and Boston Island, South Australia.

Is knowledge about species adequate for objectives and actions to be defined accurately? (If not, provide list of additional studies required): No.

- Fire regimes need to be established for some areas.

Recovery plan objectives:

- Maintain existing range and abundance.
- Reintroduce to areas of former range.
- Remove introduced population on Greenly Island due to documented changes in vegetation composition. Monitor vegetation recovery long-term.

Management actions completed or underway:

- Protection of some islands and mainland areas as reserves.
- Development and application of fire regime for Perup Nature Reserve.
- Fox control at Tutanning, Kalbarri, Dryandra, Perup.

Management actions required:

- Declaration of additional conservation reserves: yes, parts of Dryandra State Forest should be converted to Nature Reserve. Peniup Creek area should be converted to Nature Reserve.
- Habitat management: application of suitable fire regimes essential (on mainland).
- Feral animal control: fox control on mainland essential.
- Translocation: yes.
- Captive breeding: not required.
- Other: eradication of introduced population on Greenly Island, due to severe overgrazing of important island plant communities (Fatchen 1982).

Organisations responsible for conservation of species: Western Australian Department of Conservation and Land Management (CALM); South Australian National Parks and Wildlife Service (SANPWS).

Other organisation(s) or individual(s) involved: Macquarie University (Prof. D.W. Cooper); CSIRO Division of Wildlife and Ecology (W.E. Poole); Australian National Parks and Wildlife Service (ANPWS).

Conservation reserves affected by recovery plan (list action plan species in each conservation reserve): Western Australia: Perup Nature Reserve: *Dasyurus geoffroii* (chuditch), *Bettongia penicillata* (brush-tailed bettong), *Pseudocheirus occidentalis* (western ringtail), *Myrmecobius fasciatus* (numbat). Tutanning Nature Reserve: *Phascogale calura* (red-tailed phascogale), *Bettongia penicillata* (brush-tailed bettong), *Myrmecobius fasciatus* (numbat), East and West Wallabi Islands, Houtman Abrolhos. North Twin Peaks and Middle Islands, Archipelago of the Recherche. South Australia: Kangaroo Island Conservation reserves (as in *"Conservation reserves on which species occurs"* section above) and Greenly Island Conservation Park.

Other public lands affected by conservation plan: Dryandra State Forest: *Phascogale calura* (red-tailed phascogale), *Bettongia penicillata* (brush-tailed bettong), *Myrmecobius fasciatus* (numbat). Peniup Creek Reserve Western Australia. Garden Island (Commonwealth Defence Land). Granite Island (Local Government Reserve) South Australia.

Other lands affected by conservation plan: Private land near Northampton and Cranbrook Western Australia, private land on Kangaroo Island (including many areas protected by Heritage Agreements) and Boston Island, South Australia.

Can Recovery Plan be carried out with existing resources?: No.

Western Australia:

1. Prescribed fire regimes for mainland existing and proposed conservation reserves. One staff for two years—$A60,000 salaries, $A50,000 expenses.
2. Conduct necessary burning and predator control—$A20,000 per annum (x10).
3. Carry out fox control on additional reserves—$A20,000 per annum (x10).

South Australia:

1. Eradication of introduced population on Greenly Island—$A50,000.

Total=$A560,000

Species Recovery Outline No. 21

Family: Macropodidae

Scientific name: *Onychogalea fraenata*

Common names: Bridled nailtail wallaby, Merrin

Intra-specific taxa: None

Species survival status: Endangered (restricted range, locally common. Extinct from most of historical distribution).

Former distribution: From the Murray River in Victoria through New South Wales to Charters Towers in north Queensland.

Current distribution: Central Queensland west of the Great Dividing Range, Dawson/Mackenzie Rivers catchment area. Taunton Scientific Reserve, northwest of Dingo.

Habitat: Nailtails shelter by day in the edges of brigalow (*Acacia harpophylla*) scrubs and feed by night in the more open surrounding grassy eucalypt woodlands.

Reasons for decline: Effects of pastoral industry, habitat modification (proven) introduced predators (fox) and possibly competitors (rabbit). Shooting for fur and as pest species (proven, bounties paid between 1880-1916).

Conservation reserves on which species occurs: Taunton Scientific Reserve, near Dingo, Central Queensland.

Other public lands on which species occurs: None known.

Other land on which species occurs: Privately held grazing lands between Dingo and Bluff, e.g. properties adjacent to Taunton Scientific reserve.

Is knowledge about the species adequate for objectives and actions to be defined accurately? (If not, provide list of additional studies required): No.

- Interaction between bridled nailtail wallabies, black-striped wallabies, and dingoes.
- Effect of land management practices on habitat.
- To develop an effective and efficient monitoring program.
- Dispersal details of young and/or immature animals.
- Recruitment rates to the breeding populations.

Recovery plan objectives:

- Increase population's current range and abundance on Taunton Scientific Reserve through habitat enhancement.
- Increase population's range and abundance on privately held land through education and incentives.
- Establish additional captive breeding colonies and strengthen existing colonies.
- Reintroductions/relocations to suitable areas.

Management action completed or underway:

- Taunton Scientific Reserve (11,449 ha) has been gazetted for the protection of the species and its habitat.
- Management guidelines have been developed for Taunton Scientific Reserve and are being implemented. Management policies include:
  1. restricted public access
  2. removal of stock grazing
  3. controlled fire regime
  4. feral animal control
  5. some natural predator and competitor control and
  6. exotic plant eradication.
- Breeding colonies have been established at Moggill near Brisbane and at Pallarenda, Townsville.

Management actions required:

- Declaration of additional conservation reserves: yes, an expensive alternative, but may be possible. Extreme resistance from landholders. Land with suitable habitat is still available but at large cost.
- Habitat management: identification of effects of various habitat management strategies, e.g. fire. Implementation of a program to encourage management strategies on privately held lands that favour the retention of bridled nailtail wallaby habitat. Identification and protection of suitable habitat on public lands. Retention of corridors of habitat between significant areas of habitat.
- Feral animal control: implementation of an effective predator monitoring and control program, especially foxes and dingoes.
- Translocations: investigate potential areas for relocations/translocations. Research into relocation techniques.
- Captive breeding: establish additional captive breeding colonies and strengthen existing colonies.
- Other: employ a wildlife manager to implement the recovery plan. The need for integrated management of crown lands. Adoption of regional and district management regimes that involve all government and non-government agencies. The regimes must include wildlife management. Drafting and implementation of endangered species legislation to ensure identified significant habitats are protected.

Organisations responsible for conservation of species: Queensland National Parks and Wildlife Service (QNPWS).

Other organisation(s) or individual(s) involved: Australian National Parks and Wildlife Service (ANPWS).

Conservation reserves affected by recovery plan (list other action plan species in each conservation reserve): Taunton Scientific Reserve, Blacktown Tableland National Park?

Other public lands affected by the recovery plan: Some Department of Forestry lands. Minor stock routes and other utility services.

Other lands affected by recovery plan: Grazing lands held under either freehold or leasehold title in the Dingo area. Some of this land is still in its natural state and is known to support bridled nailtail wallaby populations. Survival of the species can be assured only if populations and habitat are protected on privately held land.

Can recovery plan be carried out with existing resources? (If not, list additional staff and financial resources required): No.

Note: Currently funded research program will investigate interactions between bridled nailtail wallabies, black-striped wallabies, and dingoes.

On Taunton Scientific Reserve the implementation of the management recommendations from current and past research projects is the responsibility of the Queensland National Parks and Wildlife Service. Resources are insufficient to consider populations and habitats outside the reserve.

1. Additional staff required include one wildlife manager for three years—$A120,000, one field assistant—$A90,000. Resources: vehicle—$A30,000, running expenses—$A9,000, projected related expenses—$A45,000.

   Total=$A294,000

Remarks: This species, in conjunction with others on Taunton Scientific Reserve, has considerable fundraising potential for guided nature tours. Such an operation may be considered in future years after the species becomes more secure. Any funds raised would be used for management of the Reserve.

Present and past research is management oriented. The results are being used to manage Taunton Scientific Reserve for the benefit of the species. There is a permanent management presence on the Reserve ensuring protection of the species from direct human interference and quick reaction to management crisis such as fire.

Species Recovery Outline No. 22

Family: Macropodidae

Scientific name: *Petrogale lateralis*

Common names: Black-flanked rock-wallaby, Black-footed rock-wallaby, Purple-necked rock-wallaby (Queensland)

Intra-specific taxa: *P.l. lateralis* (southwestern Cape Range and western Pilbara, Western Australia, Salisbury Island, Barrow Island, Depuch Island), Little Sandy Desert, *P.l. hackettii* (Mondrain, Wilson and Westall Islands [Archipelago of the Recherche] Western Australia), *P.l.* unnamed race (Edgar Range [western Kimberley]), *P.l.* unnamed Central Australian ranges [MacDonnell Ranges] race in Northern Territory and Western Australia, *P.l. purpureicollis* (western Queensland), *P.l. pearsoni* (South Australia, Pearson Island [Investigator Group] and Thistle and Wedge Islands [introduced]).

Species survival status: Vulnerable

Former distribution: Western Australia—from south Kimberley through the deserts and Pilbara to the wheatbelt, and islands off Pilbara and south coast. Boundary between *P.l. lateralis* and *P.l.* (central ranges race) unclear. South Australia—*P.l.* central Australian or MacDonnell Ranges race widespread and locally abundant in the MacDonnell Ranges (Northern Territory) Musgrave, Mann and Everard Ranges of northwestern South Australia and on many of the outlying hills immediately to the south. Also in ranges in central Western Australia near Northern Territory border. *P.l. pearsoni*, North Pearson Island (Investigator Group) and St. Peter Island (Nuyts Archipelago). *P.l. purpureicollis*, locally common in western Queensland.

Current distribution: Range significantly reduced and apparently still declining on mainland. Formerly widespread and abundant in its Western Australian range, but now reduced to a few remnant populations, including islands off the Australian mainland. May be extinct in the Murchison River gorge in Kalbarri National Park. Formerly widespread through the central desert region wherever isolated rocky habitat afforded suitable habitat but now reduced to the major range systems. Became extinct in Uluru National Park within last 10 years. It does not appear to have declined in Queensland.

Western Australia:
Remnant populations only in Edgar Range (? St. George Range), Calvert Range near Lake Disappointment, Little Sandy Desert, central ranges near Northern Territory border and Wheatbelt. Reasonably common in Cape Range. Extinct on Depuch Island. Abundant on Barrow, Salisbury, Mondrain, Westall, and Wilson Islands.

South Australia:
Isolated small colonies still occur in the Musgrave, Mann, and Everard Ranges but detailed survey work is necessary to determine extent of distribution and population size. North Pearson Island (163 ha) and introduced to South Pearson (39 ha—5 founders 1974), Thistle (4000 ha—10 founders 1975), and Wedge Islands (967 ha—11 founders 1975).

Northern Territory:
Mt. Olga population (Uluru National Park) extinct during the last 10 years. Most western desert populations have disappeared within the past few decades. Widespread and locally common in the central ranges.

Habitat: Rockpiles—steep cliffs and boulder-scree slopes; granite outcrops. Usually with cover of open shrubland vegetation.

Reasons for decline: Predation by foxes (proven), changed fire regimes (speculative), competition with introduced domestic and feral herbivores (speculative).

Conservation reserves on which species occurs: Western Australia: Barrow Island Nature Reserve, Cape Range and Kalbarri National Parks (possibly extinct in Kalbarri), Mount Stirling, Mount Caroline, Nangeen Hill, Tutakin, and Recherche Archipelago Nature Reserves, ten Reserves within the MacDonnell Ranges (some of them are very small) of the Northern Territory. South Australia: None. Queensland: Possibly in Lawn Hill National Park.

Other public lands on which species occurs: Vacant Crown land in Little Sandy Desert, Central Aboriginal Reserves. North and South Pearson Islands, South Australia, (to be added to Investigator Group Conservation Park). Pastoral leasehold land, Northern Territory.

Other land on which species occurs: Two populations on private land in Wheatbelt in Western Australia. Various pastoral leases and Aboriginal freehold land in the Northern Territory. Pitjantjatjara Lands (northwestern South Australia), Wedge Island and Thistle Island.

Is knowledge about species adequate for objectives and actions to be defined accurately? (If not, provide list of additional studies required): Western Australia: Yes. Northern Territory: No. South Australia: No. Queensland: No

Western Australia:
- Study of population genetics is needed.
- Fauna survey to clarify status of species in southern Kimberley region.

Northern Territory:
- Fauna survey to clarify status of species in southern part of Northern Territory.

South Australia:
- Research current distribution and status in northwestern South Australia (i.e. Pitjantjatjara Lands) in relation to rest of central Australian distribution and abundance.

Queensland:
- Population requires a distribution survey (northeastern Queensland and Winton Area), but not high priority.
- Establish program to monitor long-term changes in distribution and abundance.
- Study of basic ecology with particular reference to the role of introduced predators and herbivores in different parts of the species range, persistence/recovery of populations after environmental stress, and processes of dispersion and recolonization.

Recovery plan objectives:
- Maintain existing range and abundance.
- Halt current decline in distribution and abundance
- Increase abundance in populations that are heavily predated by foxes (selected sites).
- Reintroduce to areas of former range.
- Develop management plans for key colonies to enhance recolonization of former range and abundance through control of introduced predators and herbivores, and other management activities as appropriate.
- Research ecology of species to improve guidelines for management.

Management actions completed or underway:
Western Australia:
- Protection of much habitat in conservation reserves.
- Development and application of fox control techniques at Nangeen, Tutakin, Mount Stirling, and Mount Caroline Nature Reserves, part of Cape Range National Park.
- Survey of Central Aboriginal Reserve and Durba Hills, 1988-1990 (D.J. Pearson, in prep.).
- Determination of chromosomal race in Bell Rock Range (Eldridge et al., accepted).
- Determination of chromosomal constitution of Lake Disappointment population (Eldridge et al., in prep.)

South Australia:
- Initial partial survey of range country in northwestern South Australia (Sept. 1985) (Copley et al. 1989).
- Pearson Islands to be added to Investigator Group Conservation Park.
- Part of Wedge Island to be dedicated as conservation park.

Northern Territory:
- Implementation of fire management has commenced in the MacDonnell Ranges Reserves.

Queensland:

- Helicopter survey of Lawn Hill area.
- Chromosome survey at Lawn Hill station.

Management actions required:

- Declaration of additional conservation reserves: yes.
  Western Australia: Calvert Range of Little Sandy Desert and Edgar Range in south Kimberley.
  South Australia: Set aside areas on Wedge and Thistle Islands as conservation parks.
  Northern Territory: Enlargement of MacDonnell Ranges Reserves into comprehensive manageable system. Establishment of appropriate reserves in southwestern Northern Territory.
  Queensland: Establishment of appropriate reserves in western Queensland.

- Habitat management: yes.
  Western Australia: Fire management necessary in remnant colonies.
  Northern Territory: Expansion of fire management in MacDonnell Ranges to increase diversity in vegetation and reduce incidence of extensive wildfire.

- Feral animal control: yes.
  Western Australia: Fox control essential. Feral goat control required at Cape Range. Possible rabbit control around some central ranges populations. Feral goat and pig control required at Kalbarri National Park if population still extant.
  South Australia: Control or eradicate feral herbivores which compete with rock-wallabies. Investigate fox/cat control programs at key colonies through involvement of local Pitjantjatjara communities.

- Translocation: yes.
  Western Australia: To areas of Wheatbelt where it formerly occurred.
  Northern Territory: Re-establish species in Uluru National Park.

- Captive breeding: not required.

- Other: yes.
  Western Australia: Population genetics studies needed for long term population management. Close liaison with Aboriginal communities regarding hunting, fox baiting programs.
  South Australia: Determine current range and relative abundance. Monitor abundance at key colonies. Develop integrated management of species in South Australia, Western Australia, and Northern Territory. Implement research programs as in "Is knowledge about species adequate..." section above.

Organisations responsible for conservation of species: Western Australian Department of Conservation and Land Management (CALM); Northern Territory Conservation Commission (CCNT); South Australian National Parks and Wildlife Service (SANPWS); Queensland National Parks & Wildlife Service (QNPWS).

Other organisation(s) or individual(s) involved: Aboriginal communities of western desert; Pitjantjatjarra and Ngaanyatjarra Councils and other community councils.

Conservation reserves affected by recovery plan (list other action plan species in each conservation reserve):
Western Australia:
Barrow Island Nature Reserve: *Bettongia lesueur* (burrowing bettong), *Isoodon auratus* (golden bandicoot). Cape Range National Park, Mount Caroline, Mount Stirling, Nangeen Hill, and Tutakin Nature Reserves (Kalbarri National Park?), Salisbury, Mondrain, Wilson, and Westall Islands, Archipelago of the Recherche.

South Australia: None.

Northern Territory:
Watarrka: *Macrotis lagotis* (bilby) (reintroduced), Finke Gorge, Ormiston Gorge, Red Bank Gorge, Arltunga, Trephina, N'Dhala, Alice Springs Telegraph Station, Emily and Jessie Gaps, and Ellery Big Hole.

Queensland: None.

Other public lands affected by conservation plan: Vacant Crown land in Little Sandy Desert, Central Aboriginal Reserves. North and South Pearson Islands, South Australia,

Other land on which species occurs: Two populations on private land in Wheatbelt in Western Australia. Various pastoral leases and Aboriginal Freehold land in the Northern Territory. Pitjantjatjara Lands (northwestern South Australia), Wedge Island, and Thistle Island.

Can recovery plan be carried out with existing resources? (If not, list additional staff and financial resources required):
Western Australia: yes, except:

1. Fox control in central deserts—$A5,000 (x10) per annum.
2. Study of population genetics—$A50,000.
3. Fauna survey to clarify status of species in southern Kimberley region and Great and Little Sandy Desert—$A20,000.

South Australia: no.

1. Survey former range to determine current range and relative status—$A50,000; feral herbivore control (to be determined); fox control (to be determined—but would require ongoing commitment of resources).

Northern Territory: no.

1. Study basic ecology of species. Three-year program with two staff $A190,000 salaries, $A110,000 support.
2. Clarify distribution and status in the central desert region and establish a long term monitoring program. Two staff for first year of survey and establishment of monitoring program—$A70,000 and $A30,000 support and thereafter 4 months annually for monitoring over next 9 years—$A210,000 salaries, $A90,000 support.

Queensland: no.

1. Funding needed for survey—$A50,000

Total=$A920,000+

Remarks: Potential to coordinate monitoring with other species (i.e. *Dasycercus cristicauda* (mulgara) and *Macrotis lagotis* (bilby)) to obtain economies of scale.

Once solutions to factors causing decline obtained there are sufficient, natural populations will provide stock for translocations. No translocations should take place between ranges of subspecies.

Species Recovery Outline No. 23

Family: Macropodidae

Scientific name: *Petrogale penicillata*

Common name: Brush-tailed rock-wallaby

Intra-specific taxa: *P. p. penicillata* (parts of Great Dividing Range from about Nanango, Queensland, to Buchan and the Grampians, Victoria), *P. p. herberti* (southeastern Queensland).

Species survival status: Vulnerable (range severely reduced in Victoria, western and southern New South Wales; common in southeastern and central Queensland and upper reaches of coastal rivers in northern New South Wales). There appears to have been a decline, however, in southeastern Queensland over the past 25 years, including populations in National Parks. The population in Lamington National Park may be extinct.

Former distribution: Now absent from western slopes and plains of New South Wales except for one or two isolated locations.

Current distribution: Southeastern Australia from Rockhampton in the north to the Grampians in the south; (no longer present at Cobar or other western plains sites). Very patchily distributed in southeast New South Wales and Victoria.

Habitat: Suitable rocky areas in inland and coastal sclerophyll forests.

Reasons for decline: Foxes (proven), goats, and sheep and rabbit grazing (speculative), fires (especially frequency), and shooting.

Conservation reserves on which species occurs: Rock wallabies occur in many National Parks along the Great Divide, including Queen Mary Falls, Lamington (may be extinct), Mt. Barney and Main Range National Park in Queensland, Warrumbungles, Mt. Kaputar (now extinct?), Blue Mountains, Kanangra Boyd and Wadbilliga in New South Wales. Jenolan Caves Reserve in New South Wales. Grampians National Park and Snowy River Cobberas—Tingarinsy in Victoria.

Other public lands on which species occurs: Watagan State Forest. Rocky Range Wildlife Reserve (East Gippsland, Victoria).

Other land on which species occurs: Several areas east of Armidale, Shoalhaven River (Kangaroo Valley—private property). Coonabarabran (Warrumbungles), New South Wales.

Is knowledge about species adequate for objectives and actions to be defined accurately? (If not, list of additional studies required): Knowledge adequate for objectives only.

- Effects of exotic animals on existing colonies.
- Effects of land management practices near colonies e.g. grazing, clearing, fire regimes, fire management practices, access tracks.
- Disease rates in healthy and declining colonies e.g. parasites.
- Isolation of colonies: is distance from nearest colony a major factor in likely decline?
- Research into predator interaction (fox-dingo).
- Research into reintroduction techniques.
- Natural cycles in population numbers.
- Dispersal rates of juveniles.
- Genetic variation within and between existing Victorian populations to ascertain levels of inbreeding and long-term viability.
- Determination of habitat requirements.

Recovery plan objectives:

- Retain current range and abundance.
- Identify factors responsible for historic decline in range.
- Increase current distribution to that formerly occupied at turn of century.
- Census remaining populations in areas where species now largely absent, and identify threats to those populations.

Management actions completed or underway:

- Investigation of fertility of *penicillata/herberti* hybrids underway.
- Surveys in Grampians and East Gippsland/initial surveys.
- Fox control (Grampians).
- General preliminary work on fox control. Some site monitoring (assessment of techniques—scat size, density).
- Reports on status including management recommendations are available for both populations.
- Survey of current and former distribution in New South Wales and Victoria.
- Reintroduction to Wombeyan Caves.

Management actions required:

- Declaration of additional conservation reserves: yes, Kangaroo Valley areas (Shoalhaven River). Refuge areas in known habitat at regular intervals along Great Dividing Range. Rocky Rover area (East Gippsland). Nanango area.
- Habitat management: maintain corridors between existing colonies e.g. prevent disruption of creek lines, cliff faces, maintain bush corridors between adjacent rocky areas. Ensure that broad contact is retained between *P. p. penicillata* and *P. p. herberti*.
- Feral animal control: control goats and foxes in National Parks where remnant colonies or rock wallabies persist e.g. Warrumbungles, Grampians, Mt. Kaputar, Kanangra Boyd, Wadbilliga, Jenolan Caves. Reduce fox and rabbit populations near existing colonies (high priority). Fence small isolate colonies as experimental approach to predator control.
- Translocation: reintroductions and ongoing management/monitoring of reintroductions (i.e. to Wombeyan Caves, but should incorporate predator management). Use juveniles to recolonise neighboring, formerly occupied sites or move known family units. Studies on existing populations to assess which stocks might be used in reintroductions and adequate surveying before reintroductions are made. Strenuous efforts should be made to ensure the genetic integrity of distinct populations when undertaking such reintroduction programs.
- Captive breeding: not required. Keep colonies pure and representative of different areas along the range. Avoid wasting resources on introduced island populations (all captive colonies are currently island-originated). Zoos in Victoria should be breeding Victorian stock.
- Other: yes. Restrict 1080 poisoning of dingoes in National Parks on tablelands (e.g. Blue Mountains, Kanangra Boyd)—dingoes keep goat populations down and tend to exclude foxes (fox specificity essential). Exclude sheep and goats from around the base of rock-wallaby cliffs on private land.

Organisations responsible for conservation of species: New South Wales National Parks and Wildlife Service (NSW-NPWS); Victorian Department of Conservation & Environment (VDCE)—Bairnsdale (four or five colonies known in this region), (Fauna and Flora Division—Melbourne); Queensland National Parks and Wildlife Service (QNPWS).

Other organisation(s) or individual(s) involved: New South Wales Department of Tourism (Jenolan and Wombeyan Caves); Landholders.

Conservation reserves affected by recovery plan (list other action plan species in each conservation reserve): As per *"Conservation reserves on which species occurs"* section above. All small and medium sized native mammals that remain will benefit from fox, rabbit and goat control.

Other public lands affected by conservation plan: Watagan State Forest. Rocky Range Wildlife Reserve (East Gippsland, Victoria).

Other lands affected by conservation plan: Several areas east of Armidale, Shoalhaven River (Kangaroo Valley—private property), Coonabarabran (Warrumbungles) New South Wales.

Can recovery plan be carried out with existing resources? (If not, list additional staff and financial resources required): No.
Note: High priority is control of goats in Warrumbungles and Mt. Kaputar—may be within existing capacity of New South Wales National Parks and Wildlife Service.

Note: Many individual projects required because of the scattered distribution of colonies. Each project would have to be costed separately.

1. Development of overall recovery plan (management guidelines) for fire protection, wild-dog control, visitation to sites, etc.—$A92,000
2. Resources required for long-term monitoring of suitable sites, and additional survey and site identification and monitoring—$A70,000 per year minimum (x10).
3. Region between Shoalhaven River in New South Wales and Snowy River area in eastern Victoria must be surveyed to follow up and confirm any extant colonies (i.e. Wadbilliga). Contact between colonies is likely to be of vital importance—$A60,000 salaries, $A30,000 support.

Total: $A954,000

Remarks: The control of goats, foxes, and rabbits is an expensive and ongoing requirement for National Park managers. One shortcut method in some National Parks is to retain the dingo/wild dog population. These animals are often poisoned by the authorities to placate nearby landholders who run sheep.

Species Recovery Outline No. 24

Family: Macropodidae

Scientific name: *Petrogale persephone*

Common name: Proserpine rock-wallaby

Intra-specific taxa: None

Species survival status: Endangered (extremely restricted range).

Former distribution: ?

Current distribution: Northeast coastal Queensland, near Proserpine.

Habitat: Rainforest; closed forest/open forest interface may be important.

Reasons for decline: Unknown

Conservation reserves on which species occurs: Mt. Dryander National Park, Gloucester Island, Conway Range National Park.

Other public land on which species occurs: Possibly State Forest near Proserpine.

Other land on which species occurs: Private property near Proserpine.

Is knowledge about species adequate for objectives and actions to be defined accurately? (If not, provide list of additional studies required): No.

- Study to identify specific habitat requirements of *P. persephone* and *P. inornata*.
- Captive study of reproduction and age estimation.
- Study of social behaviour.
- Genetic diversity of and gene flow between remnant colonies.
- Study of dispersal from natal colony.
- Population demography of a number of representative colonies, especially comparison of island and mainland populations.
- Development of appropriate monitoring techniques.
- Radio telemetry studies of home range and habitat usage.

Recovery plan objectives:

- Identify range, abundance, and conservation status.
- Maintain existing range and abundance.
- Protection of known rainforest habitats (most is in national park and majority of remainder is in State Forest).
- Identify ecological requirements.

Management actions completed or underway:

- Distribution survey in progress.
- Home range and habitat usage (telemetry study due to commence).

Management actions required:

- Development of management guidelines for private property owners to protect *P. persephone* colonies.

Organisations responsible for conservation of species: Queensland National Parks and Wildlife Service (QNPWS).

Other organisation(s) or individual(s) involved: Australian National Parks and Wildlife Service (ANPWS).

Conservation reserves affected by recovery plan (list other action plan species in each conservation reserve): Mt. Dryander National Park, Gloucester Island, and Conway Range National Park.

Other public lands affected by plan: State Forests near Proserpine.

Other lands affected by plan: Private lands near Proserpine.

Can recovery plan be carried out with existing resources? (If not, list additional staff and financial resources required): No.

1. Long-term (10-year) broad-ranging research/management program incorporating home range, habitat use, dispersal, etc. Two post-doctoral scientists, two technical officers—$A120,000 per annum. Radio telemetry—$A50,000 initial set-up costs, $A5,000-$A10,000 per annum maintenance. Air charter time (for dispersal study)—$A10,000 per annum. Satellite resources—$A3,000 per transmitter + satellite—$A10,000-$A15,000 per annum. Travel, camping costs, etc.—$A20,000 per annum. Behavioural equipment (video, etc.)—$A20,000 initial cost. On-costs at 20%.

Total=$A1,763,000

Remarks: Small colonies occur in suitable habitat. There appear to be few threats to habitat.

Species Recovery Outline No. 25

Family: Burramyidae

Scientific name: *Burramys parvus*

Common name: Mountain pygmy-possum

Intra-specific taxa: None described, but the Kosciusko and Bongong populations may have differentiated.

Species survival status: Endangered

Former distribution: Fossil remains have been identified from Wombeyan Caves in southeastern New South Wales, Jenolan Caves, New South Wales and Buchan Caves, Victoria.

Current distribution: Restricted to the alpine and subalpine regions at altitudes between 1,400-2,000 m. Within this region, there are two geographically isolated populations: Mt. Bogong—Mt. Higginbotham (Victoria); Kosciusko National Park (New South Wales).

Habitat: Distribution correlates with glacial boulder-rock screes formed during last ice age. A variety of vegetation over rocks depending on rock type, depth, aspect. Most distinctive is *Podocarpus lawrencei* heathland, dominated by *P. lawrencei* and combination of other alpine and subalpine plants.

Reasons for decline: Historical and natural contraction of alpine areas. Favoured habitat may be a seral stage in vegetation development which over thousands of years may become unsuitable (speculative?). Habitat destruction through ski resort development (proven), predation by cats and foxes (proven), possible impacts associated with 100 years of vegetation modification caused by grazing and burning in alpine and subalpine areas (speculative).

Conservation reserves on which species occurs: Bogong National Park (Victoria), Kosciusko National Park (New South Wales).

Other public lands which species occurs: Most productive area known (Mt. Higginbotham—Mt. Loch, Victoria) is bisected by the Mt. Hotham Alpine Resort. Mt. Mckay site is also managed by Alpine Resorts Commission and State Electricity Commission. In Kosciusko National Park, skiing developments also impinge.

Other land on which species occurs: Not known from private land.

Is knowledge about species adequate for objectives and actions to be defined accurately? (If not, provide list of additional studies required): Yes.

Recovery plan objectives:

- Determine current distribution and abundance
- Retain current distribution.
- Determine the impact of feral predators on *Burramys* populations.
- Determine effects of ski resort developments on a sub-population at Mt. Blue Cow, New South Wales.
- Possible re-establishment in areas isolated by development and/or succession.
- Prevent further fragmentation due to man-induced changes.
- Formulate plan of management for New South Wales *Burramys* population.

Management actions completed or underway:

- Department of Conservation and Environment (VDCE) has published a comprehensive management plan, setting out management actions. Arthur Rylah Institute for Environmental Research, Technical Report Series No 66, (Ian Mansergh, Paul Kelly, and David Scotts, February, 1989).
- "Determine current distribution and abundance" and "Determine effects of ski resort developments" objectives have been or are currently being addressed.
- "Determine impact of feral predators" objective is being addressed incidentally as part of the study on Mt. Blue Cow, but further research is needed.

Management actions required:

- Declaration of additional conservation reserves: yes, management arrangements with ski resorts.
- Habitat management: prevent siltation, trampling, grazing, snow compaction, fragmentation, and burning. Greenhouse effect is identified as a major future influence on survival—will lead to habitat loss.
- Feral animal control: yes, feral cats, foxes, and dogs are known to prey on this and other small mammals of its environment.
- Translocation: yes, possible sites for reintroduction/introduction exist.
- Captive breeding: yes, in operation at Healesville Sanctuary—need research on ability of captive animals to hibernate.
- Other: continue vegetation and population monitoring over several years to evaluate long-term trends.

New South Wales management plan will seek to identify and manage all existing habitat in Kosciusko National Park, set guidelines which will avoid further fragmentation, and link adjacent *Burramys* habitat areas. Delineation of *Burramys* habitat zoning incorporated into lease amendments as required.

Organisations responsible for conservation of species: Victorian Department of Conservation and Environment (VDEC); New South Wales National Parks and Wildlife Service (NSWNPWS).

Other organisation(s) or individual(s) involved: Australian National Parks and Wildlife Service (ANPWS); Alpine Resorts Commission. Mt. Blue Cow Ski Bowl Pty. Ltd.; National Parks and Wildlife Foundation.

Conservation reserves affected by recovery plan (list other action plan species in each conservation reserve): Bogong National Park, Victoria, Kosciusko National Park, New South Wales.

Other public lands affected by plan: Alpine Ski Resorts.

Other lands affected by plan: None.

Can recovery plan be carried out with existing resources? (If not, list additional staff and financial resources required): Yes (Victoria), but funding is required in New South Wales.

1. Ongoing monitoring program of species at Mt. Blue Cow in Kosciusko National Park as a necessary addendum to current study being conducted by the New South Wales National Parks and Wildlife Service 1990-1992, 1998-2000. Personnel (three weeks field work plus one week collate results and write reports), $A7,000 (x10).
2. Expansion of monitoring program to cover all *B. parvus* populations in Kosciusko $A7,000 (x10).
3. Preparation of management plan—$A5,000, implementation, $A100,000 per annum (x10).

Total=$A1,145,000

Remarks: Work has been carried out by Ian Mansergh as partial fulfillment of Ph.D. requirements as employee of VDCE. Also work by David Scotts—two years research as exempt employee of VDCE. Vast database in existence—approximately eight years of data. Research funded by Victorian Dept. Conservation and Environment, World Wide Fund for Nature Australia, and Australian National Parks and Wildlife Service.

The cost and logistics of predator control as a means of determining the impact of feral predators on *Burramys* populations cannot be estimated at present. Further securing of habitat is required within resort areas through appropriate zoning and lease amendments to ensure full protection of habitat.

A management plan will be forthcoming from New South Wales National Parks and Wildlife Service for that state's population upon completion of a current (1991) study by Linda Broome of the effects of ski resort development on the *Burramys* population of Mt. Blue Cow in Kosiusko National Park.

Possibly require scientist to investigate implications of Greenhouse effect on future of species.

Species Recovery Outline No. 26

Family: Petauridae

Scientific name: *Pseudocheirus occidentalis*

Common name: Western ringtail

Intra-specific taxa: None

Species survival status: Vulnerable

Former distribution: Restricted to the semi-arid and wetter parts of southwestern Australia from near Perth to east of Albany.

Current distribution: Isolated populations in coastal areas over its former range and some populations in inland jarrah forests.

Habitat: Prefers areas where peppermint *Agonis flexuosa* is either dominant or an understory species, but may also be found in paperbark thickets (*Melaleuca*), *Casuarina roman,* and *Eucalyptus* spp. forests and woodlands.

Reasons for decline: Probably a combination of predation pressure (foxes), fragmentation of habitat (agriculture and urbanization), and increased fire frequency (speculative).

Conservation reserves on which species occurs: Two Peoples Bay Nature Reserve, Leeuwin-Naturaliste National Park, Perup Nature Reserve, Tuart National Park.

Other public lands on which species occurs: Ludlow State Forest and several other near-coastal forest areas.

Other land on which species occurs: Coastal stands of *Agonis* adjacent to Geographe Bay and east of Augusta. Townships of Busselton, Dunsborough, Quindalup, and Albany.

Is knowledge about species adequate for objectives and actions to be defined accurately? (If not, provide list of additional studies required): No.

Recovery plan objectives: None devised yet.

Management actions completed or underway:

- Attempts to obtain funding for basic research. Preliminary study underway by Western Australian Museum.

Management actions required:

- Declaration of additional conservation reserves: yes, possibly after suitable knowledge of distribution and habitat requirement is obtained.
- Habitat management: prevention of regular wildfire.
- Feral animal control: fox control required.
- Translocation: reintroduction into areas where the species previously occurred as determined by old dreys (nests) or after the requisite habitat evaluation.
- Captive breeding: possibly, but conservation measures in the wild may be adequate after species biology is understood.
- Other: evaluation of the systematics of the disjunct populations using electrophoresis or DNA fingerprinting.

Organisations responsible for conservation of species: Western Australian Department of Conservation and Land Management Department (CALM).

Other organisation(s) or individual(s) involved: Western Australian Museum (Ric How); Australian National Parks and Wildlife Service (ANPWS).

Conservation reserves affected by recovery plan (list other action plan species in each conservation reserve): Plan not yet devised.

Other public land affected by conservation plan:

Other lands affected by conservation plan:

Can recovery plan be carried out with existing resources? (If not, list additional staff and financial resources required): No.

1. Detailed documentation of distribution and habitat preferences. Assessment of genetic variation in population over species range. Assessment of basic biological parameters e.g. breeding, diet, density, longevity. Preparation of recovery plan—$A200,000 over three years.

Total=$A200,000

Species Recovery Outline No. 27

Family: Petauridae

Scientific name: *Gymnobelideus leadbeateri*

Common name: Leadbeater's possum

Intra-specific taxa: None

Species survival status: Endangered (uncommon, restricted). Population estimate (1985) 5000. Predictions are that, regardless of action, populations *will* decline substantially. However, survival is probable given appropriate future forest management.

Former distribution: Previously thought extinct until rediscovery in 1961 in Victoria's Central Highlands range; always highly restricted. Previous range also included areas south and northeast of present distribution. Fossils from New South Wales.

Current distribution: Victoria: Central Highlands. Decline of 50-90% predicted.

Habitat: Moist tall open forest, dominated by montane ash (mountain ash, alpine ash, shining gum), much of which is theoretically available for harvesting. Species needs nest sites in living or dead old-growth trees and foraging areas in regrowth ash and wattle. Species also has relatively narrow time envelope in forest succession. Nest sites are presently at a premium, will become even more so as dead trees fall or if old-growth trees are cut.

Reasons for (predicted) decline: Inappropriate forestry practices (destruction of old-growth forests), loss of suitable nesting trees (proven). Current major threat to species is both fragmentation of habitat leading to small non-viable population isolates and high rate of tree collapse of existing hollow-bearing trees—lack of nest sites for the species for the next 150 years (proven).

Conservation reserves on which species occurs: Baw Baw National Park, Lake Mountain State Park (no significant occurrences in reserves).

Other public lands on which species occurs: Most of the public forests supporting ash-type trees in the central highlands, i.e. state forests, water catchment areas, Yellingbo State Forest Reserve.

Other land on which species occurs: None.

Is knowledge about species adequate for objectives and actions to be defined accurately? (If not, provide list of additional studies required): No.

- Introduce silvicultural techniques which allow some trees to eventually become suitable nesting sites for the species.
- Effects on isolated colonies ability to disperse.
- Evaluate effectiveness of retained unlogged corridors in logged forest for maintenance of viable populations, and for animal dispersal between isolates.
- Determine minimum population sizes for short- and long-term survival in isolated patches.
- Predict greenhouse effects on habitat fragmentation and isolation.

Recovery plan objectives:

- Maintain and expand current distribution and population levels in state forests and water catchment areas.
- Identify distribution of suitable habitat.
- Identify effectiveness of reserves.
- Modify current logging practice to ensure maintenance of Leadbeater's possum within logged coupes (rather than in unlogged areas) where possible.
- Determine sensitive silvicultural techniques for ash-type forests.
- Design and establish a system of well-managed nature reserves dedicated to Leadbeater's possum protection and management.
- Maintain captive breeding program.
- Reintroduce the species to former range and additional forest isolates (very low priority).

Management actions completed or underway:

- Draft management plan has been written which aims to provide forest management strategies and prescriptions (M.A. MacFarlane and J.H. Seebeck, Feb. 1991). Guidelines derived from draft plan are being implemented.
- Certain actions already in place in regions—mapping changes of harvesting prescriptions, roading.

Management actions required:

- Declaration of additional conservation reserves: yes, identify, map, and protect refuge areas (multi-age ash forest) that will provide tree hollows for the next 150 years. Reservation and protection of all remaining mature and multi-aged ash forest in the central highlands (approx 5,000 ha in scattered isolates) in special purpose reserves. Design a viable reserve system that will be suitable in the long term, especially under impacts of the greenhouse effect.
- Habitat management: Reduce impacts of logging practices on species by increasing number of hollow bearing trees available to the species and protect existing nest tree resource; specifying silvicultured management to create suitable habitat; modifying current Silviculture Systems Project design to incorporate studies of the effects of logging coupe pattern (e.g. retention of unlogged strips of forest and around coupes) on survival and habitat use by species; continuing ban on logging in water catchment areas.
- Feral animal control: Not required.
- Translocation: Not required.
- Captive Breeding: Maintain and improve current captive breeding capacity and techniques.
- Other: Establishment of a management committee. Prepare a Geographic Information System database for Central Highlands ash forests at a scale 1:25000 for assessing species' total population sizes; habitat distribution, fragmentation and isolation; for mapping pattern and location of unlogged corridors and refuge systems and for modelling and monitoring future population changes due to habitat succession and greenhouse effects. Establish a program for monitoring changes in distribution and abundance. List all of species' habitat on the Register of the National Estate. Public education programs.

Organisations responsible for conservation of species: Victorian Department of Conservation and Environment (VDCE).

Other organisation(s) or individual(s) involved: David Lindenmayer (Arthur Rylah Institute); Andrew Smith (University of New England); Timber Industry; La Trobe University (Neil Murray); Healesville Sanctuary; Melbourne Zoo (Chandi De Alwis).

Conservation reserves affected by recovery plan (list other action plan species in each conservation reserve): Baw Baw National Parks, Lake Mountain State Park. All hollow-dependent fauna would benefit from improvements in tree hollow retention in state forest—all arboreal marsupials in ash forest.

Other public lands affected by plan: All state forests and water catchment areas in species current range.

Other lands affected by plan: None.

Can recovery plan be carried out with existing resources? (If not, list additional staff and financial resources required): No.
Note: Department of Conservation and Environment allocated $A200,000 for 1989-1990 for research and management plan implementation. Costs include loss of timber production (although not excessive) and staff resources for effective implementation of management plan when completed.

1. GIS establishment approximately $A100,000 (but note Department of Conservation and Environment has established some GIS data already and more is planned).
2. Monitoring costs—potentially low if organised by existing regional staff using volunteer possum watchers—$A15,000 (x10).

Total= $A250,000+

Remarks: As populations of Leadbeater's possum will be dispersed, fragmented, and possibly isolated, it will be essential to determine the dispersal capabilities of the species.

Appendix 3: Action Plan Authors and Australasian Marsupial and Monotreme Specialist Group Members (1988-1990)

Associate Professor M. Archer
School of Zoology
University of New South Wales
P.O. Box 1
Kensington NSW 2033
AUSTRALIA

Mr. P.R. Brown
Dept. Science
Victoria College
Rusden Campus
662 Blackburn Road
Clayton VIC 3168
AUSTRALIA

Dr. A.A. Burbidge
Director of Research
Dept. Conservation and Land Management
P.O. Box 51
Wanneroo WA 6065
AUSTRALIA

Dr. J. Calaby
c/o Commonwealth Scientific and Industrial Research Organisation (CSIRO) Division of Wildlife and Ecology
P.O. Box 84
Lyneham ACT 2602
AUSTRALIA

Mr. P. Copley
S.A. National Parks & Wildlife Service
G.P.O. Box 1782
Adelaide SA 5001
AUSTRALIA

Dr. M. Denny (Chairperson)
Mt. King Ecological Surveys
'Oorong' Bathurst Road
Oberon NSW 2787
AUSTRALIA

Ms. Joan Dixon
Curator of Mammals
Museum of Victoria
328 Swanson Street
Melbourne Vic 3000
AUSTRALIA

Dr. T. Flannery
Australian Museum
P.O. Box A285
Sydney South NSW 2000
AUSTRALIA

Mr. G. George
Department of Science
Victoria College-Rusden Campus
662 Blackburn Road
Clayton VIC 3168
AUSTRALIA

Dr. J. Giles
Assistant Director
Taronga Park Zoological Gardens
P.O. Box 20
Mosman NSW 2088
AUSTRALIA

Dr. G. Gordon
Senior Zoologist
Qld National Parks & Wildlife Service
P.O. Box 42
Kenmore QLD 4069
AUSTRALIA

Dr. K.A. Johnson
Conservation Commission of the Northern Territory
P.O. Box 1046
Alice Springs NT 0871
AUSTRALIA

Dr. C. Kemper
Curator of Mammals
Museum of South Australia
North Terrace
Adelaide SA 5000
AUSTRALIA

Mr. M. Kennedy
(Secretary & Action Plan Compiler)
Senior Project Officer
World Wide Fund for Nature Australia (WWFA)
G.P.O. Box 528
Sydney NSW 2001
AUSTRALIA

Mr. G.R. Kula
Assistant Secretary
Dept. Environment and Conservation
P.O. Box 5266
Boroko
PAPUA NEW GUINEA

Dr. L. Lim
21/32 Gerrard St
Cremorne NSW 2096
AUSTRALIA

Dr. G. Maynes
Manager
Wildlife Monitoring Unit
Australian National Parks and Wildlife Service (ANPWS)
P.O. Box 636
Canberra ACT 2601
AUSTRALIA

Dr. S. Morton
Principal Research Scientist
Commonwealth Scientific and Industrial Research Organisation (CSIRO)
Centre for Arid Zone Research
P.O. Box 2111
Alice Springs NT 0871
AUSTRALIA

Mr. Stephen Nash
Director
Trade Records Analysis of Flora and Fauna in Commerce (TRAFFIC) (Southeast Asia)
P.O. Box 10769
50724 Kuala Lumpur
MALAYSIA

Dr. A.E. Newsome
Senior Principal Research Scientist
Commonwealth Scientific and Industrial Research Organisation (CSIRO)
Division of Wildlife and Ecology
P.O. Box 84
Lyneham ACT 2602
AUSTRALIA

Mr. D. Papps
Deputy Director (Policy and Wildlife)
NSW National Parks & Wildlife Service
P.O. Box 1967
Hurstville NSW 2220
AUSTRALIA

Mr. R. Pearse
Assistant Director
Dept of Parks, Wildlife and Heritage
G.P.O. Box 44A
Hobart TAS 7001
AUSTRALIA

Mr. W.E. Poole
Principal Research Scientist
Commonwealth Scientific and Industrial Research Organisation (CSIRO)
Division of Wildlife and Ecology
P.O. Box 84
Lyneham ACT 2602
AUSTRALIA

Dr. T. Press
Australian National Parks and Wildlife Service
Kakadu National Park
P.O. Box 71
Jabiru NT 0886
AUSTRALIA

Mr. Widodo S. Ramono
Sub-Directorate of Species Monitoring
Directorate General of Forest Protection
and Nature Conservation
15 Jalan Ir H. Juanda
Bogor 16122
INDONESIA

Prof. Barry Richardson
Foundation Chair-Biological Sciences
University of Western Sydney-Hawkesbury
Burke St.
Richmond, NSW 2753
AUSTRALIA

Dr. W.D.L. Ride
Australian National University
G.P.O. Box 4
Canberra ACT 2601
AUSTRALIA

Mr. A.C. Robinson
Senior Survey & Research Officer
SA National Parks & Wildlife Service
G.P.O. Box 1782
Adelaide SA 5001
AUSTRALIA

Mr. David Rounsevell
Department of Parks, Wildlife & Heritage
G.P.O. Box 44A
Hobart TAS 7001
AUSTRALIA

Mr. J.H. Seebeck
Arthur Rylah Institute for
Environmental Research
Dept. Conservation Forests and Lands
P.O. Box 137
Heidelberg Vic 3084
AUSTRALIA

Mr. L. Seri
Senior Ecologist
Dept. Environment and Conservation
Nature Conservation Division
P.O. Box 5266
Boroko
PAPUA NEW GUINEA

Dr. A. Smith
Senior Lecturer
Dept. Ecosystem Management
University of New England
Armidale NSW 2351
AUSTRALIA

Mr. R. Strahan
Photographic Index
Australian Museum
P.O. Box A285
Sydney South NSW 2000
AUSTRALIA

Dr. Peter Teirney
QLD National Parks & Wildlife Service
P.O. Box 904
Emerald QLD 4720
AUSTRALIA

Mr. R.M. Warneke
Arthur Rylah Institute for
Environmental Research
Dept. of Conservation and Environment
P.O. Box 137
Heidelburg VIC 3084
AUSTRALIA

Dr. C.H.S. Watts
Museum of South Australia
North Terrace
Adelaide SA 5000
AUSTRALIA

Dr. G. Wilson
Department of Primary Industry & Energy
Bureau of Rural Resources
Brisbane Avenue
Barton ACT 2600
AUSTRALIA

Dr. J.W. Winter
P.O. Box 151
Ravenshoe QLD 4872
AUSTRALIA

Appendix 4: List of Tables and Figures

Tables

Figures

ACIUCN. 1989. *The Conservation of Threatened Species and Their Habitats.* Proceedings of a national conference, Sydney, 1987. Australian Committee for IUCN (Occasional Papers 2). Sydney.

Aitken, P.F. 1983. Sandhill Dunnart. In: *The Australian Museum Complete Book of Australian Mammals,* ed. by R. Strahan. Angus and Robertson, Syney, pp. 60.

Allendorf, F.W., and Leary, R.F. 1986. Heterozygosity and fitness in natural populations of animals. In: *Conservation biology: the science of scarcity and diversity,* ed. by M. E. Soulé. Sinauer Associates: Sunderland, Massachusetts. Pp. 57-76.

Aplin, K.P. 1987. Basicranial anatomy of the early Miocene diprotodontan *Wynyardia bassiana* (Marsupialia: Wynyardiidae) and its implications for wynyardiid phylogeny and classification. In: *Possums and Opossums: Studies in Evolution,* ed. by M. Archer. Royal Zoological Society of New South Wales: Sydney. Pp. 369-91.

Anon. 1982. Mammals, reptiles, butterflies and fish protected in Indonesia. In: *Traffic Bulletin,* 3(6), pp. 63-65.

—. 1988. *Laporan Tahunan Balai Konservasi Sumber Daya Alam VIII.* Departemen Kehutanan, Direktorat Jenderal Perlindungan dan Pelestarian Alam.

ANPWS. 1989. *An Australian National Strategy for the Conservation of Species and Habitats Threatened with Extinction—A draft for public comment.* Australian National Parks and Wildlife Service, Canberra.

APAG. 1990. *The Barnett Report—A Summary of the Report of the Commission of Inquiry into Aspects of the Timber Industry in Papua New Guinea.* The Asia-Pacific Action Group. Hobart.

Archer, M. 1981. Systematic revision of the marsupial genus *Sminthopsis* Thomas. *Bull. Amer. Mus. Nat. Hist.* 168:61-224.

Baker, R.M., and George, G.G. 1988. Species management programmes in Australia and New Zealand. *Int. Zoo Yearb.* 27:19-26.

Ballou, J.D., and Foose, T.J. (in press). In: *Wild mammals in captivity,* ed. by S. Lumpkin and D. Kleiman. University of Chicago Press: Chicago.

Baynes, A. 1987. The original mammal fauna of the Nullarbor and southern peripheral regions: Evidence from skeletal remains in superficial cave deposits. In: *A biological survey of the Nullarbor region, South and Western Australia,* ed. by N.L. McKenzie and A.C. Robinson. Government Printer, Adelaide. Pp. 139-52.

Brown, P.R. 1989. *Management Plan for the Conservation of the Eastern Barred Bandicoot, (Perameles gunnii),* In: *Victoria,* ed. by J.H. Seebeck. Arthur Rylah Institute for Environmental Research, Technical Report Series No. 63.

Burbidge, A.A. 1989. Conservation values of Western Australian islands. In: *Australian and New Zealand islands: nature conservation values and management,* ed. by A.A. Burbidge, Occasional Paper No. 2/89. Department of Conservation and Land Management, Perth.

Burbidge, A.A. and Jenkins, R.W.G. (eds.) 1984. *Endangered vertebrates of Australia and its island Territories.* Australian National Parks and Wildlife Service, Canberra.

Burbidge, A.A. and McKenzie, N.L. 1989. Patterns in the modern decline of Western Australia's vertebrate fauna: causes and conservation implications. *Biological Conservation* 50:143-198.

Burbidge, A.A., Johnson, K.A., Fuller, P.J. and Southgate, R.I. 1988. Aboriginal knowledge of the mammals of the central deserts of Australia. *Australian Wildlife Research* 15:9-39.

Burton, R., Kennedy, M., and Fry, I. 1986. The threatened vertebrates. In: *A Threatened Species Conservation Strategy for Australia—Policies for the Future,* ed. by M. Kennedy and R. Burton. Ecofund Australia, Sydney. Pp. 8-12.

Burton, J.A., and Pearson, B. 1987. *Collins Guide to the Rare Mammals of the World.* William Collins, London.

Calaby, J.H. 1971.The current status of Australian Macropodidae. *Australian Zoologist* 16:17-29.

Callister, D. 1990. *Trade in Australian Marsupials and Monotremes.* A report prepared for the IUCN/SSC Australasian Marsupial and Monotreme Specialist Group. Unpublished.

Clark, T.W., Backhouse, G.N. and Lacy, R.C. 1991. Report of a workshop on population viability assessment as a tool for threatened species management and conservation. *Aust. Zool.* 27:28-35.

Clark, T.W. and Seebeck, J.H. 1990. *Management and Conservation of Small Populations.* Proceedings of a Conference held in Melbourne, Australia, September 26-27, 1989. Chicago Zoological Society, Brookfield, Illinois.

Clark, T.W., Warneke, R.M., and George, G.G. 1990. Management and conservation of small populations. In: *Management and conservation of small populations,* ed. by T.W. Clark and J.H. Seebeck. Chicago Zoological Society: Brookfield, U.S.A. Pp. 1-18.

Coman, B.J. 1983. Fox *Vulpes vulpes.* In: *The Australian Museum Complete Book of Australian Mammals,* ed. by R. Strahan. Angus and Robertson, Sydney. Pp. 486-73.

CONCOM. 1991. *List of Australian Endangered Vertebrate Fauna* (April 1991). ANPWS, Canberra.

Copley, P.B., Kemper, C.M. and Medlin G.C. 1989. The mammals on north-western South Australia. *Rec. S. Aust. Mus.* 23:75-88.

Costin, A.B., and Groves, R.H. (eds.) 1973. *Nature Conservation in the South Pacific.* Proceedings of Symposium A-10 XII Pacific Science Congress, Canberra, 1971. IUCN Publications New Series No. 25.

Craven, I., and de Fretes, Y. 1987. *The Arfak Mountains nature conservation area, Irian Jaya: management plan 1988-1992.* WWF Project 3770.

Delroy, L.B., and Earl, J., Radbone, I., Robinson, A. C. and Hewett, M. 1986. The breeding and re-establishment of the Brush-tailed Bettong, *Bettongia penicillata* in South Australia. *Aust. Wildl. Res.* 13: 387-396.

Diamond, J.M. (1984). Historic extinctions: A rosetta stone for understanding prehistoric extinctions. In: *Quaternary Extinctions,* ed. by P.S. Martin and R.G. Klein. University of Arizona Press, Tucson. Pp. 824-62.

—. Undated. *A Proposed Natural Reserve System for Papua New Guinea.* Physiology Dept, University of California Medical Centre

Diamond, J. 1986. *Proposed studies of conservation problems on the Western Papuan islands (Raja Ampat islands), Irian Jaya, Indonesia.* File report to WWF Irian Jaya.

Eaton, P. 1985a. Land Tenure and Conservation: Protected Areas in the South Pacific, Noumea. In: *South Pacific Regional Environmental Programme Topic Review* No. 17. Pp. 49-58.

—. 1985b. Protected Areas in the South Pacific; 1. Tonda Wildlife Management Area. *Ples.* Pp. 4-14.

—. 1986. *Grassroots Conservation—Wildlife Management Areas in Papua New Guinea.* Land Studies Centre Report S6/1—University of Papua New Guinea.

Elliot, H.F.I. 1973. Past, Present and Future Status of Pacific Islands. In: *Nature Conservation in the South Pacific,* ed. by A.B. Costin and R.H. Groves. Australian National University. Canberra. Pp. 217-227.

FAO. 1981. *National Conservation Plan for Indonesia: Vol. VII, Maluku and Irian Jaya.* FAO, Bogor.

Fatchen, T.J. 1982. Vegetation distribution and change of offshore islands of the Investigator Group and Whidbey Isles, Great Australian Bight. *Trans. R. Soc. S. Aust.* 106:39-60.

Flannery, T.F. 1990. *Mammals of New Guinea.* The Australian Museum. Robert Brown and Associates: Queensland.

Flesness, N.R. and Foose, T.J. 1990. The Role of Captive Breeding in the Conservation of Species. Guest Essay in 1990 IUCN Red List of Threatened Animals, compiled by the World Conservation Monitoring Center and IUCN—The World Conservation Union.

Fletcher, M. and Maynes, G. 1988. Macropod Research Needs and Priorities. ANPWS Draft Discussion Paper, Canberra.

Foose, T.J. 1983. The relevance of captive propagation to the conservation of biotic diversity. In: *Genetics and Conservation: a reference for managing wild populations,* ed. by C.M. Schonewald-Cox, S. M. Chambers, B. MacBride and W. L. Thomas. Benjamin/Cummings: Menlo Park, California. Pp. 374-401.

Foose, T.J., Lande, R., Flesness, N. R., Rabb, G. and Read, B. 1986. Propagation plans. *Zoo Biology.* 5(2): 139-146.

Fosberg, F.R. 1973. Past, Present and Future Conservation Problems of Oceanic Islands. In: *Nature Conservation in the Pacific,* ed. by A.B. Costin and R.H. Groves, Australian National University Press, Canberra. Pp. 209-215.

Frankel, O.H. and Soulé, M.E. 1981. *Conservation and Evolution.* Cambridge University Press, Cambridge.

Franklin, I.R. 1980. Evolutionary change in small populations. In: *Conservation biology: an evolutionary-ecological perspective,* ed. by M.E. Soulé and B.A. Wilcox. Sinauer associates: Sunderland, Massachusetts. Pp. 135-150.

George, G.G. 1979. The status of endangered Papua New Guinea mammals. In: *The status of endangered Australasian Wildlife,* ed. by M. Tyler, Royal Zoological Society of South Australia, Adelaide. Pp. 93-100.

—. 1990. Monotreme and marsupial breeding programs in Australian Zoos. *Aust. J. Zool.* 37: 181-205.

—. 1990. Captive breeding of endangered species. In: *Australia's Endangered Species,* ed. by M. Kennedy. Simon and Schuster Australia: Brookvale, NSW. Pp.153-159.

• George, G.G., Dixon, J., Challis, G. and Lacy, R.C. 1990. The taxonomy and palaeontology of the Eastern Barred Bandicoot: their bearing on management of the Victorian population. In: *The management and conservation of small populations,* ed. by T.W. Clark and J.H. Seebeck. Chicago Zoological Society: Brookfield, Illinois.

George, G.G., and Maynes, G.M. 1990. The Status of New Guinea Bandicoots. In: *Bandicoots and Bilbies,* ed. by J.H. Seebeck, P.R. Brown, R. Wallis and C. Kemper. The Australian Mammal Society/Surrey Beatty and Sons: Sydney. Pp 93-105.

Giles, J.R. and Lim, L. 1987. Conservation of the endangered native species of New South Wales: an assessment of current status, threats and requirements for maintenance of wildlife populations. In: *The Conservation of Threatened Species and Their Habitats,* ed. by M. Hicks and P. Eiser, Australian Committee for IUCN, Occasional Papers 2. Pp. 90-100.

Gilpin, M.E. 1987. Spatial structure and population vulnerability. In: *Viable populations for conservation,* ed. by M. E. Soulé. Cambridge University Press: Cambridge. Pp. 125-139.

Gilpin, M.E. and Soulé, M.E. 1986. Minimum viable populations: processes of species extinction. In: *Conservation biology: the science of scarcity and diversity,* ed. by M.E. Soulé. Sinauer Associates: Sunderland, Massachusetts. Pp. 17-34.

Gressitt, J.L. and Ziegler, A.C. 1973. The effects of the loss of Forest in New Guinea. In: *Nature Conservation in the Pacific,* ed. by A.B. Costin and R.H. Groves. Australian National University Press, Canberra. Pp. 117-122.

Gressitt, J.L. and Gagne. 1982. Conservation in New Guinea. In: *Biogeography and Ecology of Papua New Guinea,* ed. by J.L. Gressit, W. Junk, The Hague.

Groves, C.P., and Flannery, T.F. 1990. Higher level systematics within the Peramelidae (Marsupialia). In: *Bandicoots and Bilbies,* ed. by J.S. Seebeck, P.R. Brown, R. Wallis and C. Kemper. The Australian Mammal Society/Surrey Beatty and Sons: Sydney.

Hide, R.H., Pernetta, J.C. and Senabe, T. 1984. Utilisation of wild animals. In: *South Simbu: Studies in demography, nutrition and subsitence,* ed. by R.H. Hide. Research Report 6, Simbu Land use Project. PNG Institute of Applied Social and Economic Research. Pp. 291-379.

Hill, R. 1991. A Research Recovery Plan for the Brush-tailed Rock-wallaby *Petrogale penicillata penicillata* (Gray 1825), in South-eastern Australia. (Unpublished report to ANPWS Endangered Species Program; Department of Conservation and Environment, Victoria).

IUCN. 1988. Captive breeding—the IUCN policy statement. *Species* 10: 27-28.

—. 1990. *1990 IUCN Red List of Threatened Animals.* IUCN World Conservation Monitoring Centre, Cambridge, UK.

Johnson, K.A., Burbidge, Andrew A. and McKenzie, Norman L. 1989. Australian Macropodoidea: causes of decline and future research and management. In: *Kangaroos, Wallabies and Rat-Kangaroos,* ed. by G.C. Grigg, P.J. Jarman and I.D. Hume. Australian Mammal Society, Sydney. Pp. 641-657.

Kimber, R. 1983. Black lightning: Aborigines and fire in central Australia and the western desert. *Archaeology in Oceania* 18:35-48.

King, B. 1989. Does wildlife management by the people work? A case study of the Pokili Wildlife Management Area, Papua New Guinea. *Science in New Guinea* 15(8).

Kinnear, J.E., Onus, M.L. and Bromilow, R.N. 1988. Fox control and rock wallaby population dynamics. *Australian Wildlife Research* 15:435-447.

Kirkpatrick, I., Meredith, C., Norton, T., Plumwood, V. and Fenshaw, R. 1990. *The Ecological Future of Australia's Forests,* ed. by D. McLlroy. Australian Conservation Foundation, Melbourne.

Kitchener, D.J., Chapman, A., Muir, B.G. and Palmer, M. 1980. The conservation value for mammals of reserves in the Western Australian Wheatbelt. *Biol. Conserv.* 18:179-207.

• Kitchener, D.J. 1991. *Pseudontechinus mimulus* (Thomas, 1906) (Marsupalia, Dasyuridae): rediscovery and description. *Records of the Western Australian Museum* 15:191-202.

Koopman, K.F. 1979. Zoogeography of Mammals from Islands off the North Eastern Coast of New Guinea. *Am. Mus. Novit.* 2690:1-16.

Lacy, R.C. 1987. Loss of genetic diversity from managed populations: interacting effects of drift, mutation, immigration, selection and population subdivision. *Cons. Biol.* 1:143-58.

—. 1989. Analysis of founder representation in pedigrees: founder equivalents and founder genome equivalents. *Zoo Biol.* 8:111-24.

Lacy, R.C., and Clark, T.W. 1990. Population viability assessment of the Eastern Barred Bandicoot in Victoria. In: *The management and conservation of small populations,* ed. by T.W. Clark and J.H. Seebeck. Chicago Zoological Society: Brookfield, Illinois. Pp. 131-146.

Lacy, R.C., Flesness, N.R., and Seal, U. 1989. *Puerto Rican Parrot population viability analysis.* Report to the U.S. Fish and Wildlife Service. Captive Breeding Specialist Group, Species Survival Commission, IUCN: Apple Valley, Minnesota.

Lamb, K.P. and Grass, J.L. (eds.) 1976. *Ecology and Conservation in Papua New Guinea.* Wau Ecology Institute Pamphlet No. 2. PNG.

Lande, R., and Barrowclough, G.F. 1987. Effective population size, genetic variation, and their use in population management. In: *Viable populations for conservation,* ed. by M.E. Soulé. Cambridge University Press: Cambridge. Pp. 87-123.

Latz, P.K. and Griffin, G.F. 1978. Changes in Aboriginal land management in relation to fire and to food plants in central Australia. In: *The nutrition of Aborigines in relation to the ecosystem of central Australia,* ed. by B.S. Hetzel and H.J. Frith. CSIRO, Melbourne. Pp. 77-85.

Laurie, E.M.O., and Hill, J.E. 1954. *List of land mammals of New Guinea, Celebes and Adjacent Islands* 1758-1952. British Museum (Natural History): London.

Lim, L., Robinson, A. C., Copley, P. B., Gordon, G., Canty, P. D. and Reimer, D. 1987. *The Conservation and Management of the Yellow-footed Rock-wallaby, (Petrogale xanthopus), Gray, 1854*. National Parks and Wildlife Service, Department of Environment and Planning, South Australia. Spec. Publ. 1987. No. 4.

Low, W.A. and Strong, B.W. 1987. Mammals of Lake Nash and Georgina Downs Stations and a First Record of the Narrow-nosed Planigale *Planigale tenuirostris* from the Northern Territory. *Northern Territory Naturalist* 10:3-5.

Lunney, D., and Leary, T. 1988. The impact on native mammals of land-use changes and exotic species in the Bega district, New South Wales, since settlement. *Australian Journal of Ecology* 13:67-92.

MacFarlane, M.A., and Seebeck, J.H. 1991. Draft Management Strategies for the Conservation of Leadbeater's Possum *Gymnobolideus leadbeateri*, in Victoria. Arthur Rylah Institute for Environmental Research, Technical Report Series No. 111.

Maguire L.A., Seal, U.S., and Brussard, P.F. 1987. Managing critically endangered species: the Sumatran rhino as a case study. In: *Viable populations for conservation*, ed. by M.E. Soulé. Cambridge University Press: Cambridge. Pp. 141-158.

Maguire, L.A., Lacy, R.C., Begg, R.J., and Clark, T.W. 1990. An analysis of alternative strategies for recovering the Eastern Barred Bandicoot in Victoria. In: *Management and conservation of small populations*, ed. by T. W. Clark and J. H. Seebeck. Chicago Zoological Society: Brookfield, Illinois. Pp. 147-164.

Mansergh, I., Kelly, P. and Scotts, D. 1989. *Management Strategy and Guidelines for the Conservation of the Mountain Pygmy Possum, (Burramys parvus) in Victoria*. Arthur Rylah Institute for Environmental Research, Technical Report Series No. 66.

Martin, R. 1989. *Draft Management Plan for the Conservation of the Koala (Phascolarctos cinereus) in Victoria*. Arthur Rylah Institute for Environmental Research, Technical Report Series No. 99.

Menzies, J.I. 1976. Some thoughts on the ecology and conservation of mammals, reptiles and amphibians. In: *Ecology and Conservation in Papua New Guinea*. Proceedings of Wau Ecology Institute Symposium, 4th Nov. 1975, ed by K. P. Lamb and J.L. Gressit. Wau Ecology Institute Pamphlet No. 2.

Menzies, J.I. and Pernetta, J.C. 1986. A taxonomic revision of the cuscusses allied to *Phalanger orientalis* (Marsupialia: Phalangeridae). *J. Zool., Lond.* (B) 1:551-618.

Mitchell, A.H. 1988. *Implementation of conservation in Irian Jaya, Indonesia; progress report for March-June 1988*. WWF/IUCN Project 3770.

—. 1987. *Progress in establishment of Irian Jaya's system of conservation areas—report no. 2*. WWF/IUCN Project 3770.

Morton, S.R. 1990. The impact of European settlement on the vertebrate animals of arid Australia: a conceptual model. *Proceedings of the Ecological Society of Australia*, ed. by D.A. Saunders, A.J.M. Hopkins and R.A. How. 16:201-213.

Morton, S.R. and Baynes, A. 1985. Small mammal assemblages in arid Australia. *Aust Mammal.* 8 159-69.

Nash, S.V. 1989. *Progress report for April-June 1989*. WWF Project 3770.

—. 1989. *Progress report for July-September 1989*. WWF Project 3770.

Native Mammal Taxon Advisory Group. 1991. Unpublished report to the Conservation Co-ordinator from Native Mammal Taxon Co-ordinator on Taxon Advisory Group Meeting, Western Plains Zoo, Dubbo.

Newsome, A.E. 1983. Dingo *Canis familiaris dingo*. In: *The Australian Museum complete book of Australian mammals*, ed. by R. Strahan. Angus and Robertson, Sydney. Pp. 483.

Nicholson, P.H. 1981. Fire and the Australian Aborigine—an enigma. In: *Fire and Australian biota*, ed. by A.M. Gill, R.H. Groves and I.R. Noble. Australian Academy of Science, Canberra.

Norton, T.W. and Saxon, M.J. 1991. *Management of the Long-footed Potoroo (Potorous longipes) in South-east New South Wales*. Cres, ANU, NSW NPWS (unpublished).

Papua New Guinea. 1975. *The Constitution of the Independent State of Papua New Guinea*. Port Moresby, PNG.

Patchett, M.A. 1991. *Development of Proposals for a Major Involvement in Papua New Guinea by WWF*. Consultancy Report to World Wide Fund For Nature South Pacific Program. Sydney.

Pernetta, J.C., and Hill, L. 1983. The impact of traditional harvesting on endangered species: the Papua New Guinea experience. In: *Endangered Species: Social, Scientific, Economic and Legal Aspects in Australia and the South Pacific*, ed. by Shelly Burgin, Total Environment Centre. Pp. 98-130.

—. 1982. *International Pressure on Internal Resources Management in Papua New Guinea*. Monograph 16. Institute of Applied Social and Economic Research. PNG. Pp. 319.

—. 1984. Traditional Land Use and Conservation of Resources in the Pacific Basin. *Ambio* 13:329-364.

Pearson, D.J., and Robinson, A.C. 1990. New Records of the Sandhill Dunnart, *Smithopsis psammophila* (*MARSUPIALIA: DASYURIDAE*) in South Western Australia. *Aust. Mammal.* 13:57-59.

Petocz, R. 1984. *Conservation and development in Irian Jaya: a strategy for rational resource utilization*. WWF/IUCN Project 1528/PT Sinar Agape Press.

—. 1983. *Recommended reserves for Irian Jaya province*. WWF/IUCN Project 1528 special report.

Petocz, R. and de Fretes, Y. 1983. *Mammals of the reserves in Irian Jaya*. WWF/IUCN Project 1528 special report.

Ralls, K. and Ballou, J.D. 1983. Extinction: lessons from zoos. In: *Genetics and Conservation: a reference for managing wild populations*, ed. by C.M. Schonewald-Cox, S. M. Chambers, B. MacBride and W.L. Thomas. Benjamin/Cummings: Menlo Park, California. Pp. 164-184.

Ratcliffe, J.B. 1984. *Cagar alam Pegunungan Cyclops, Irian Jaya, Indonesia: Management Plan 1985-1989*. WWF/IUCN Project 1528.

Recher, H.F., and Lim, L. 1990. A review of current ideas of the extinction, conservation and management of Australia's terrestrial vertebrate fauna. *Proceedings of the Ecological Society of Australia*, ed. by D.A. Saunders, A.J.M. Hopkins and R.A. How. 16:287-301.

Ride, W.D.L. and Wilson, G.R. 1982. The conservation status of Australian mammals. In: *Species at Risk: Research in Australia*, ed. by R. H. Groves and W. D. L. Ride. Australian Academy of Sciences, Canberra. Pp. 27-45.

Robinson, N.A., Sherwin, W.B., Murray, N.D. and Graves, J.A.M. 1990. Applications of conservation genetics to the Eastern Barred Bandicoot, *Perameles gunnii*. In: *The management and conservation of small populations*, ed. by T.W. Clark and J.H. Seebeck. Chicago Zoological Society: Brookfield, Illinois. Pp. 109-129.

Robinson, A.C. and Young, M.C. 1983. The Toolache Wallaby *Macropus greyi* Waterhouse. *Department of Environment and Planning Special Publications No. 2*. Department of Environment and Planning, Adelaide.

Schodde, R. 1973. General Problems of Fauna Conservation in Relation to the Conservation of Vegetation in New Guinea. In: *Nature Conservation in the South Pacific*, ed. by A.B. Costin and R.H. Groves. Australian National University, Canberra. Pp. 123-144.

Schultze-Westrum, T.G. 1971. Conservation in Papua New Guinea, Final Report on the 1979 World Wildlife Fund Mission, WWF Switzerland.

Scotts, D. and Seebeck, J.H. 1989. Ecology of *Potorous longipes* (Marsupialia: Potoroidae); and preliminary recommendations for management of its habitat in Victoria. Arthur Rylah Institute for Environmental Research, Technical Report Series No. 62.

Seal, U., and Lacy, R.C. 1989. *Florida Panther population viability analysis.* Report to the U. S. Fish and Wildlife Service. Captive Breeding Specialist Group, Species Survival Commission, IUCN: Apple Valley, Minnesota.

Shaffer, M.L. 1981. Minimum population sizes for species conservation. *Bioscience* 31:131-34.

—. 1987. Minimum viable populations:coping with uncertainty. In: *Viable populations for conservation,* ed. by M. E. Soulé. Cambridge University Press: Cambridge. Pp. 69-86.

Sherwin, W.B. and Murray, N.D. 1989. Population and Conservation Genetics of Marsupials. *Aust. J. Zool.*

Soulé, M.E. 1980. Thresholds for survival: maintaining fitness and evolutionary potential. In: *Conservation biology: an evolutionary-ecological perspective,* ed. by M.E. Soulé and B.A. Wilcox. Sinauer associates: Sunderland, Massachusetts. Pp. 151-170.

Soulé, M., Gilpin, M., Conway, W. and Foose, T. 1986. The millenium ark: how long a voyage, how many staterooms, how many passengers? *Zoo Biology* 5:101-113.

Strahan, R. 1989. *Australian Museum Complete Book of Australian Mammals.* 2nd printing. Angus and Robertson Publishers, Sydney.

Strahan, R. Unpublished MS. "Survival Status of Australian Marsupials".

Thompson, E.A. In press. Genetic importance and genomic descent. In: *Population Management for survival and recovery,* ed. by J.D. Ballou, T.J. Foose, and M.E. Gilpin. University of Chicago Press: Chicago.

Total Environment Centre. 1983. *Our Wildlife in Peril.* A.H. and A.W. Reed Australia, Sydney.

•van Der Zon, A.P.M. 1979. *Mammals of Indonesia.* FAO, Bogor.

van Dyck, S.M. 1991. Raising an old glider's ghost. *Wildlife Australia* Winter, 1991. Pp. 10-13.

Walton, D.W. (ed.) 1988. *Zoological Catalogue of Australia, 5, Mammalia.* Australian Government Publishing Service, Canberra.

Watts, C.H.S. (ed.) 1991. *A list of vertebrates of South Australia.* Biological Survey Co-ordinating Committee/Department of Planning and Environment, South Australia, Adelaide.

Whitten, A.J., Mustafa, M., and Henderson, G.S. 1988. *The Ecology of Sulawesi.* Gadjah Mada University Press, Yogyakarta.

World Bank. 1990. *Papua New Guinea The Forestry Sector: A Tropical Forestry Action Plan Review Report.*

WWF Indonesia. 1978. *Nature conservation in Irian Jaya.* WWF Indonesia Programme.